Persuading Science

The Art of
Scientific Rhetoric

Persuading Science

The Art of Scientific Rhetoric

Marcello Pera
and
William R. Shea

Editors

Science History Publications, USA

1991

First published in the United States of America
by Science History Publications/U.S.A.
a division of
Watson Publishing International
Post Office Box 493
Canton, MA 02021

Library of Congress Cataloging-in-Publication Data

Persuading science : the art of scientific rhetoric / Marcello
Pera and William R. Shea, editors.
    p. cm.
  Includes bibliographical references and index.
  ISBN 0-88135-071-0
1. Science—Philosophy.   2. Persuasion (Rhetoric)
I. Pera, Marcello, 1943–   .   II. Shea, William R.
Q175.Ps84   1991
501—dc20   91-10554
                                                        CIP

Designed and Manufactured in the U.S.A.

Contents

Part II   Rhetoric in Action

Preface

It is common knowledge that the Scientific Revolution was accompanied by a rejection of scholasticism and the use of rhetoric in investigating nature. This is stressed in many contemporary works but perhaps nowhere with greater jubilation over the "exantlation of truth" than in Thomas Sprat's *History of the Royal Society*, where we are told that the members of the new Society set out "to separate the knowledge of Nature from the colours of Rhetorick, the devices of Fancy, or the delightful deceit of Fables". Their remedy "for this extravagance", Sprat adds, was the

> constant Resolution, to reject all the amplifications, digressions, and swellings of style: to return back to the primitive purity, and shortness, when men deliver'd so many things, almost in an equal number of words. They have exacted from all their members, a close, naked, natural way of speaking; positive expressions, clear senses; a native easiness; bringing all things as near the Mathematical plainness, as they can: and preferring the language of Artizans, Countrymen, and Merchants, before that, of Wits, or Scholars.[1]

On the Continent, scientists like Galileo inveighed against the endless quibbling of academics and begged them to realize that one does not study nature in the same way one pours over the pages of an epic poem like the *Iliad* or the *Odyssey*.[2] Roughly at the same time, Descartes made it clear that "ordinary dialectic is of no use whatever to those who wish to investigate the truth of things. Its sole advantage is that it sometimes enables us to explain to others arguments which are already known. It should therefore be transferred from philosophy to rhetoric."[3]

Without going into the merits of this protracted polemic, we cannot forget that it played an important role in the genesis of modern science. None of the essays in this volume wishes to query the fact or the originality of the Scientific Revolution, but all of them, from widely different viewpoints, examine the reasons why the campaign against rhetoric was so often conducted with literary panache and more than average rhetorical skill. The central question that is raised in this volume is not whether science gives us genuine knowledge but how it justifies the knowledge that it acquires, transforms and disseminates. According to the image of science that prevailed until fairly recently, two facets had to be examined. The first may be called *epistemological* and can be described as the belief that the foundation of knowledge is twofold: it rests namely on (a) the *facts* that are disclosed in observation and experiments, and (b) the abstract *ideas* that govern mathematical reasoning and the rules of logical inference. The second aspect is *methodological* and covers the cognitive strategies that science has developed to get to the facts or to justify the theories that bring the facts under laws that explain why, where and how facts occur. What is striking about this broad and necessarily vague characterization of science is that it has two poles and two poles only: the investigating mind and the investigated nature. Nature supplied the perceptions, mind the conceptions. When the perception, the observation or the experimental result were untainted, the conception was assumed to be unbiased, and the knowledge objective.

This image of science was substantially altered in the last few decades. On the scientific side, new theories displaced older ones that were considered unassailable. On the philosophical side, the two poles of the scientific image came under review. What has been called the dogma of "immaculate perception" was subjected to scrutiny, and the alleged rock-bottom of hard facts lost much of its consistency. The vaunted objective method of science was found to be less detached than advertised. The result was a strong reaction against science. This was not new in itself since science has often been the butt of criticism from various sources that claimed that it was, for instance, inferior to religious insight or aesthetic intuition. What was novel was the claim that science did not provide genuine knowledge and that its methods and its results were mere social conventions. As Paul Feyerabend put it: "Scientists, being embedded in constantly changing social surroundings, used ideas (and, much later, equipment) to *manufacture* first metaphysical atoms, then crude physical atoms, then complex systems of elementary particles out of a material that did not contain these elements but could be shaped by them."[4] Without sharing Feyerabend's radical historicism and while retaining the view that science does provide us with genuine knowledge, the essays in this volume recognize that our image of science has to be revised. As far as the epistemological aspect of the question is concerned, there is no fundamental disagreement, and all are agreed that we do not have access to pure facts, and that all observation is, to some extent, theory-laden. The debate centers on the *meth-*

odological issue, and, if the authors of this book are united in repudiating the allegation that there is no scientific method, they vary in their characterization of the proper method or methods. Philip Kitcher, Marcello Pera, Paolo Rossi and Dudley Shapere stress that the demise of the naive view that there is a unique and infallible scientific method does not entail that all scientific knowledge is void of rational justification. But what kind of rational justification? The answer is sought in an examination of the rhetoric of science, which would seem to land us in a veritable mine-field since the word rhetoric can, by itself, set off a number of explosive reactions. After Hegel and Marx, the word "dialectics" is also open to suspicion. The more neutral term "persuasion" can be used as long as we bear in mind that it involves the presentation of arguments to an audience (namely the use of rhetoric) and the refutation of opponents (which belongs to dialectics).

The essays in this book are divided into two sections. The first analyzes why rhetoric is relevant to scientific discourse, the second how rhetoric was actually used in famous historical cases. The first reason why science uses rhetoric is the need that every original thinker experiences when he wishes to formulate and convey a new idea. He must not only convince himself that he is right, he must also anticipate the obstacles that he will encounter when he tries to convince others. Philip Kitcher shows that this was Darwin's situation. Earlier on it was the situation of Galileo and Descartes, who had to create their own audience, as Westfall explains. Gerald Holton provides modern instances of the rhetoric of assertion by means of which a scientist convinces himself, and the rhetoric of appropriation or rejection that is used by those whom he is trying to convince. This dialectical exchange, when it does not degenerate into sterile disputation, enhances the quality of the arguments proposed. Rhetoric is an instrument of communication and a tool to remove obstacles to communication.

But there is to be a deeper reason for claiming that rhetoric is relevant to scientific discourse. It follows from the realization that science is not merely a collection of observations and that competing theories can cover the same identical facts. As Ernan McMullin points out, scientists have to go beyond facts to shared values or, as Gerald Holton expresses it, the community of scientists invariably turns to general themata or world-views that determine what is to be deemed relevant or adequate. In this perspective, rhetoric is more than an ornament; it is the way cognitive claims are discussed and assessed. It serves not only to transmit information but to acquire and validate knowledge. This does not mean that we can ignore the basis from which the rhetorical argument proceeds, rather, as Marcello Pera argues, it invites us to examine the foundation of the rhetorical strategy.

The result is a modification of our image of science. Where we had two protagonists we now find three; nature and the individual researcher are joined by the community. The framework is expanded from a two-dimensional plane bounded by nature and the individual researcher to a three-dimensional manifold that includes the wider community of scientists and scholars. Rhetoric is the instrument that furthers the continuous exchange between individuals and the broader community to which they belong. It provides the rules of the game. In a profound sense, it broadens the traditional idea we have of the scientific method. Yet we must bear in mind that rhetoric is localized. Argument begins with the "'final-form' language of science", as Dudley Shapere says, namely from a shared body of ascertained facts, common results, accepted techniques, and approved theories. By accepting the role that rhetoric plays in science, we abandon a certain triumphalist historiography that saw progress as the application of a rigorously objective method of enquiry. In a sense, we are returning to Aristotle who believed that in science, as in other fields of human knowledge, we must be aware of all available means of persuasion and use them wisely. Such an approach will enrich our interpretation of the Scientific Revolution, for the dramatic shift that occurred in the seventeenth century was not merely the substitution of one tradition by another but a complex intermingling of viewpoints. Paolo Rossi shows how the Baconian method, so stridently Anti-Aristotelian, is intimately linked to traditional rhetoric. Peter Machamer suggests that the whole science of the seventeenth-century is based on a principle dear to Protagoras. Maurizio Mamiani reveals how Newton's "rules of philosophizing" come straight out of traditional logic, and William Shea discloses how Descartes ushered rhetoric out of the front door only to allow it back through a rear window in order to convince himself as well as others that his cosmology was right.

M.P. & W.R.S., 1991

NOTES

1. Thomas Sprat, *The History of the Royal Society*. London, 1667; facsimile with introduction and notes by J.I. Cope and H.W. Jones, London: Routledge and Kegan Paul, 1959, p. 62.

2. Galileo Galilei, *Il Saggiatore* in A. Favaro (ed.), *Le Opere di Galileo Galilei*, 20 vols. Florence: Barbèra 1890-1909, vol. VI, p. 232.

3. René Descartes, Rule 10 of the *Rules for the Direction of the Mind* in *The Philosophical Writings of Descartes*, translated by J. Cottingham, R. Stoothoff and D. Murdoch, 2 vols. Cambridge: Cambridge University Press, 1985, vol. I, 37.

4. Paul Feyerabend, "Realism and the Historicity of Knowledge" in William R. Shea and Antonio Spadafora (eds.), *Creativity in the Arts and Science*. Canton, MA: Science History Publications, 1990, p. 151.

Acknowledgment

The editors are grateful to the Istituto Italiano per gli Studi Filosofici of Naples for providing the intellectual and financial resources that made this book possible, and they wish to express their special thanks to the President, Gerardo Marotta, and the Secretary General, Antonio Gargano, for their invaluable assistance.

PART I

Science and Persuasion

Persuasion

PHILIP KITCHER

I yet beseech your Majesty,
If for I want that glib and oily art
To speak and purpose not, since what I well intend
I'll do't before I speak, that you make known
It is no vicious blot, murder, or foulness,
No unchaste action or dishonored step,
That hath deprived me of your grace and favor;
But even for want of that for which I am much richer,
A still-soliciting eye, and such a tongue
That I am glad I have not, though not to have it
Hath lost me in your liking.[1]

So Cordelia, offering a traditional view of rhetoric. The function of the "glib and oily art" is to disguise the thoughts, plans, intentions, and reasons of one person (the *source*), thereby influencing the thoughts, decisions, and actions of another (the *target*). In the context of discussing science, there is a canonical situation. The target comes to form a representation about nature—believing a particular proposition or accepting the accuracy of a particular diagram, let us say —on the basis of verbal performances by the source that do not present logically cogent reasons for accepting that representation. The deceptive function of rhetoric here is to give the semblance of convincing reasoning where no such reasoning exists.

This widely accepted view thrives on distinguishing ways in which people can be brought to belief. At one extreme, subjects are threatened, coerced, drugged, deceived. At the other, they are prompted by the words of others to undergo for themselves logically cogent reasoning. Rhetoric belongs to the former category of cases, figuring among the modes of deception. Hence the felt need in scientific

3

prose for an austere idiom, one in which nature and reason can be trusted to speak for themselves.

I shall argue that this view needs rethinking. Starting from a highly idealized picture of human reasoning, it draws a sharp distinction where there is a continuum of cases. To recover that continuum, I recommend that we scrutinize the phenomenon of persuasion, and so reclaim the possibility of valuable functions of rhetoric.[2]

I. A Simple-Minded View of Cognitive Subjects

Persuasive situations are naturally thought of in terms of source and target, but persuasion and rhetoric also play a role in our dialogues with ourselves. Later, I shall explore the implications of the idea that scientists might persuade themselves through acts of writing or thinking, and that the words they use on these occasions might have important rhetorical functions. For the present, however, I shall take a persuasive situation to be constituted by two individuals, the source and the target. As a consequence of the source's words, the target undergoes a psychological process culminating in the formation of a belief. We may consider two standards for appraising what occurs:

> *Perfect communication.* The process induced in the target is isomorphic to that which sustains the homologous belief in the source.
>
> *Good outcomes.* The process induced in the target belongs to a type that has a high frequency of generating true beliefs.

Although Cordelia's conception of rhetoric emphasizes its interference with the transparence of language—seeing verbal performance properly as a device for conveying the internal states of one to the mind of another—the *Good Outcomes* standard is, I believe, more fundamental. For what we should really care about is the means by which people can be led to form beliefs in ways that are epistemically virtuous.

In many instances, both criteria seem to be satisfied. The most obvious examples come from mathematics texts. Consider the following proof of the elementary theorem that the angles of a triangle sum to 180°.

> Consider any triangle ABC. Draw CD parallel to AB, and extend BC to BE. Angles ACB, ACD, DCE sum to 180°. By theorems on parallel lines, ACD = BAC, DCE = ABC. Hence ACB + BAC + ABC = 180°. Q.E.D.

In this instance, the authors aim to reproduce in the minds of readers the same sequence of inferential transitions that underlie their own beliefs in the theorem,

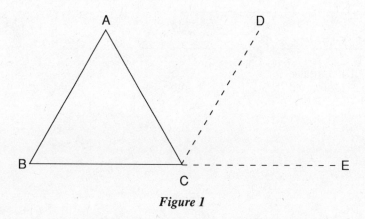

Figure 1

and thereby to enable those readers to undergo processes of a highly reliable type.[3]

Let us scrutinize more closely this conception of the target as undergoing a reliable process (or one that is isomorphic to the process sustaining belief in the source). How should we conceive of these processes? The simplest idea is to think in terms of a silent tokening of sentences: our subjects say to themselves, "ACD = BAC," then "DCE = ABC," then "ACB + BAC + ABC = 180°." But a moment's reflection suggests that the sequence of mental tokenings is not enough for what sources typically hope to inculcate. The aim is to move through that sequence as a result of making inferential transitions. So we need to think of subjects as having at their command propensities for moving from one statement (or set of statements) to another. When the subjects read or hear the proof, these propensities are activated, so that the transition to the next step is overdetermined, both by the application of the propensity and through the registration of the next part of text.

I shall embed this propensity-activation conception of following a proof— or, more generally, an argument—in a picture of our cognitive life that emerges in recent work in cognitive science (most explicitly in the writings of John Anderson) and that accords with our everyday ideas about our mental operations. Think of subjects as having a relatively small "screen" on whose contents they can focus their attention. Call this "working memory." Ordinary perception, reading texts, listening to the speech of others, all introduce statements into working memory. Other statements, including declarative statements and expressions of goals, can be brought into working memory from two much more extensive storage systems, *declarative memory* and *goal memory*. Another storage system contains the pro-

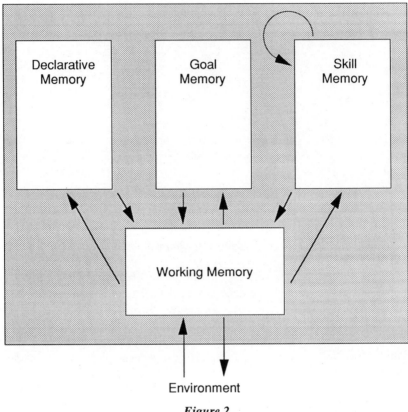

Figure 2

pensities, including but by no means exhausted by inferential propensities, and activation of these propensities can result in their application to the contents of working memory. Call this storage system *skill memory*.[4] Items from working memory can be deposited in any of the three long-term storage systems. Moreover, efforts to retrieve items from the storage systems may be successful, totally unsuccessful, or may bring unwanted items in train with those for which search was initiated. The overall picture is presented in Figure 2.

Now let us return to our simple proof. The sequence of sentences that constitute the proof prompts a complex process in the minds of readers: propensities are activated to generate images of triangles and to extend those images; results about parallel lines are recalled from declarative memory; inferential propensities are activated, so that, for example, the presence of "ACD = BAC," "DCE = ABC," and "ACB + ACD + DCE = 180°" in working memory yields the presence of "ACB + BAC + ABC = 180°" in working memory.

I hope that this reconstruction of what occurs in following the simple proof will seem so obvious as to be pedantic and unnecessary. If we take it seriously, we can begin to understand what happens in the all-too-common instances where we *fail* to follow a cogent proof. Perhaps we lack the propensities that enable us to achieve the basic conceptualization of the problem (indicated in the complaint "I don't see why things get set up in this way"); perhaps we cannot retrieve from declarative memory anything corresponding to some crucial line ("Where did that come from?"); or perhaps we fail to activate an inferential propensity that enables us to make a transition ("How does that follow?"). These common reactions to a proof are reminders that *even in the most austere case, namely, mathematics*, a rhetorical function is served by the presentation of the proof. Given two presentations of the same proof, one may be effective in prompting members of a particular audience to undergo the intended/reliable process of belief formation, while the other may not.

Relativization to audiences is crucial here. A proof presentation that is effective for one audience (professional mathematicians, say) can be useless for others (students, mathematophobes). Sometimes mathematicians and philosophers suggest that there is a canonical form in which genuine proofs *ought* to be presented. So, for example, it might be claimed that real proofs of arithmetic identities are those generated in first-order formalizations or even in first-order versions of set theory without abbreviatory conventions. Or, to use Hardy's conception of proofs, we might think of a proof as a sequence of unadorned formulae whose connections can be recognized by the mathematical expert, with all commentary dismissed as "gas."[5] Both suggestions fail for the same reason. A monstrous first-order derivation of the theorem of the infinity of primes from unabbreviated set theory would be unsurveyable and would in consequence place such strain on human working memory that it would be incomprehensible. Even if one is more liberal, allowing for inferential transitions that the cognoscenti can easily reconstruct and deploying lemmas that are familiar to them, some proofs will be too long and complicated to serve their epistemic functions without explicit commentary. It is no accident that proof strategies for Gödel's incompleteness theorem or for the Feit-Thompson theorem are explicitly outlined in advance.

"Gas" is necessary even in professional mathematics. The use of such phrases as "The strategy of the following is . . . ," "We proceed by proving a sequence of preliminary lemmas . . . ," "The essential ideas of the proof are akin to those deployed in . . . 's construction of . . . " signals the need to accommodate human cognitive limitations. If the target is to be induced to carry out the right epistemic performances, those limitations have to be acknowledged by clothing the bare logical skeleton.

II. "Dead" Rhetoric

What does this have to do with rhetoric in science? Plenty. I have deliberately begun with the hardest case, mathematics, because I want to illustrate a general point. A crucial part of persuasion in science as in everyday life is the formulation of thoughts and reasons in ways that will enable the limited cognitive systems that are the targets of persuasion to think well about the pertinent issue. In mathematics the rhetorical forms are both spare and conventionalized. In consequence mathematics typically repeats a small number of pieces of dead rhetoric, devices so commonly used that they become the equivalents of cliché.[6] But these modes of persuasion are continuous with the more vivid examples that we shall consider later. There is no line to be drawn between the devices that focus the attention of professional mathematicians on the structures of complex proofs and the images deployed by scientific revolutionaries attempting to make compelling a new vision of the order of nature.

Dead rhetoric is not confined to mathematics. It flourishes in the experimental reports that dominate scientific literature. Consider one of the simplest types, in which the experimenter suggests that a regularity of a particular kind holds. We can divide the persuasive task into two parts: convincing the target that the assignment of measurements to the quantities that figure in the sample is correct, and defending the representativeness of the sample. Discharging both tasks requires framing the presentation in ways that will anticipate and block potential challenges. Within subdisciplines such potential challenges are often so predictable that anticipation of them is mechanical. Experimenters duly record that they measured X, made sure that Y was held constant, and varied the sample with respect to Z. Rhetorical effectiveness demands knowledge of which elements need to be mentioned, which can be omitted. Frequently, although not always, the lists are so stable that the rhetoric is effectively dead.

It is worth looking a little more closely at the persuasive functions of experimental description. Steven Shapin has argued (persuasively!) that the original motivation for detailed experimental descriptions among the seventeenth-century "experimental philosophers" (for example, the prolix Robert Boyle) was to make the readers into "virtual witnesses" of the phenomena.[7] The official rationale for "methods sections" in contemporary research reports is different: they serve to enable readers to replicate the experiments. The actual function of these descriptions connects with both suggested goals, while remaining distinct from them. No scientist thinks that readers will find their descriptions so vivid as to convey the impression that the reader is really peering over the experimenter's shoulder. Scientists and students of science know all too well that the information provided in "methods sections" is typically inadequate for replication of the experiment. Very often, extended verbal descriptions prove insufficient and the would-be replicator must acquire the needed skills from people who have already done the experiment. However, as Boyle and his contemporaries hoped, descriptions can often make replication unnecessary. They persuade audiences that the work has been properly done by showing that the experimenter was attuned to the right kinds of things. The "methods description" is a display, announcing to readers that the investigator knows the business of performing this type of experiment. The rhetorical function of the details about how various quantities were measured or how the specimens were prepared is to testify to the credentials of the writer. In effect, what is said is this: "I know that this collection of results could be misleading in various ways, and I have responded to the potential pitfalls that we all know about. You can trust me."

Dead rhetoric begins to flicker into life when there is debate about whether some variable not on the standard roster might prove relevant. Most people who have listened to scientific presentations can recall exchanges that fit the following pattern:

> *Questioner*: But did you check the viscosity? Dumblebunny and Nitwit did some studies in which they showed that changes in viscosity can sometimes make a big difference.
>
> *Experimenter*: Yes, I know about the Dumblebunny-Nitwit study. But, as you are probably aware, Halfbrain was unable to replicate the viscosity effect. We did some pilot studies in which we varied the viscosity, and it made no difference.

Whether traces of such exchanges emerge in the description of the experiment depends on the experimenter's perception of community-wide attitudes. If there are only a handful of people who give any credence to the Dumblebunny-Nitwit findings, then the "methods section" will remain silent about matters of viscosity. As support for Dumblebunny and Nitwit is perceived to grow, different kinds of rhetorical device will be used to anticipate and block the potential challenge: "Since Halfbrain has shown that the Dumblebunny-Nitwit effect is not robust, measurements of viscosity were not made." "There has been considerable controversy about possible effects of changes in viscosity. Our pilot studies reveal no such effect in the present case." And so forth.

Persuasive communication of inductive argument takes the following general form: Source announces a description of the results obtained with respect to a sample and a generalization of those results to the entire population. Target critically appraises this conclusion by searching declarative memory for lists of those factors likely to interfere with proper generation of results for the sample and of those properties that a representative sample should exemplify. So a checklist is drawn up in working memory, and the conclusion is accepted when each item on the list is accounted for. Source succeeds in persuading target just in case source anticipates the contents of the lists that target's search of declarative memory will produce and provides descriptions of how these items have been accommodated.

III. Uncharted Territory

Everyday reading of research reports calls upon skills that are acquired in scientific apprenticeship and that generate predictable modes of critical appraisal. To persuade others requires no great rhetorical creativity, but simply the ability to understand how readers will need to be soothed and reassured. As we move beyond those examples in which the space of possibilities is clear and well understood by the participants, rhetoric becomes crucial in avoiding miscommunication and in focusing the targets' attention on relevant features of the case.

I shall consider a famous example in which new explanations of familiar phenomena are defended on the grounds that those explanations enjoy virtues over their rivals.[8] The source defends a position that is widely rejected as absurd, even dangerous. As a result, persuasion is only possible if numerous rhetorical problems can be overcome: the apparent difficulties of the position must be identified; they must either be addressed, or the potential for overcoming them must be exhibited; the potential dangers of the position must be defused; the virtues of the position and the defects of its rivals highlighted. In solving these rhetorical problems, subsidiary difficulties arise. So, for example, in delineating the difficulties of rivals, it is necessary to counter potential objections to the effect that the most promising developments of those rivals have been ignored.

Darwin's *Origin of Species* is a marvelously persuasive document, one that surprises many of my students by its readability and vividness.[9] The *Origin* bids to establish several distinct conclusions, some of them nearer to Darwin's heart than others, and I shall concentrate on that small set of related claims about which Darwin was most convincing. By 1867 both he and Huxley were prepared to announce victory for the thesis that species are related by descent with modification and that some questions in comparative anatomy, biogeography, paleontology, and embryology can be answered by articulating those histories of descent. Darwin's argument for these contentions consisted in first presenting evidence for the possibility of a mechanism of change in the traits of members of a species, namely, natural selection; second, in arguing that apparent objections to the possibility of modification across species boundaries could be overcome; and third, in demonstrating how many puzzling phenomena of natural history could be explained in a systematic and unified fashion if organisms were related by descent with modification. This organization of his case was not explicit in the *Origin*, although it emerges quite clearly both in Darwin's letters and in the debates of the 1860s. Within this general scheme, Darwin assembled numerous bits of detail— "facts" as he called them—to support the main ideas. Appealing to the experiences of plant and animal breeders, he stressed the extent to which organisms can be modified by conscious selection. Drawing on details of natural history, he tried to show that problems about the emergence of complex organs or about the absence of transitional forms in the fossil record are not as insuperable as they might initially appear. Finally, he offered again and again instances from the organic world that are, as he put it, "inexplicable on any other hypothesis." [10]

Consider now the targets—Darwin's readers, and more specifically, Darwin's British readers. They had to reconstruct the line of argument I have sketched from Darwin's discursive presentation. Some of them found this difficult, others impossible. William Hopkins, mathematician and physicist, lamented the absence of general principles and complained that the broadcasting of details confused the mind.[11] Even Huxley conceded that the *Origin* is a hard book, "a sort of intellectual pemmican."[12] For Darwin's former shipmate, Captain Robert Fitzroy, the act of reading the *Origin* was apparently too painful. At the famous British Association meeting in 1860, he stalked the hall, wielding a Bible and intoning "The Book! The Book!"

Let us survey some of the problems Darwin faced in trying to persuade his British contemporaries:

1. dismissal because of perceived threat to religion;
2. failure to recognize the structure of the argument;
3. protests that selection can't transcend species boundaries;
4. evaluation of some problem (e.g., the incompleteness of the fossil record) as insuperable;
5. refusal to grant that descent with modification would provide the best explanation of phenomena in natural history.

I shall explore these problems by using our simple model of human cognition and show how specific pieces of Darwinian rhetoric are designed to overcome them.

Problems 3–5 arise because of a generic difficulty with nondeductive inference. It is possible for a source to formulate a cogent argument with true premises, and for target to reject that argument by adding a further true premise. Suppose I argue for the conclusion that England will win its World Cup[13] match against Egypt by advancing correct information about the relevant qualities of the two teams.

You are not convinced because you have heard a true report that Waddle has pulled a muscle. Unbeknownst to you, Waddle has been recovering very fast and is fully fit to play in the match. Part of the rhetorical difficulty in presenting nondeductive arguments is to forestall the addition of partial truths that would undermine those arguments.

Source must thus anticipate the kinds of items that target might introduce from declarative memory that would activate propensities issuing in rejection of the intended conclusions. Consider how this is done in presenting Darwin's argument. It will not be enough simply to document instances in which breeders have modified, say, pigeons in dramatic ways. For recognizing that Darwin hopes to show the possibility of modifications across species boundaries, target may formulate the demand that some of the examples should reveal such modifications. Readers may also articulate the concern that the selective power of nature is weaker than that of the conscious breeder. To block these objections, Darwin needs to emphasize the extent of time over which natural selection can operate, and the ability of selection to scrutinize the "minute parts" of organisms. In the following passage both ideas are brought together in a sequence of striking images:

> It may be said that natural selection is daily and hourly scrutinizing, throughout the world, every variation, even the slightest; rejecting that which is bad, preserving and adding up all that is good; silently and insensibly working, whenever and wherever opportunity offers, at the improvement of each organic being in relation to its organic and inorganic conditions of life. We see nothing of these slow changes in progress, until the hand of time has marked the long lapse of ages, and then so imperfect is our view into long past geological ages, that we only see that the forms of life are now different from what they formerly were.[14]

So the objection that recorded history shows no change across species boundaries—perhaps by appeal to the famous mummified crocodiles that Geoffroy brought back from Egypt—is muffled by providing a vision of the history of life that undermines the expectation that history should show such changes.

In similar fashion Darwin's images provide a context in which apparently large problems for his views will come to seem less severe. The chapter on "The Imperfection of the Geological Record" is a rhetorical masterpiece, attempting to confront "the most obvious and gravest objection which can be urged against my theory."[15] Darwin begins with a succinct statement of the problem:

> But just in proportion as this process of extermination has acted on an enormous scale, so must the number of intermediate varieties, which

have formerly existed on the earth, be truly enormous. Why then is not every geological formation and every intermediate stratum full of such intermediate links?[16]

The question is to be answered by tutoring naive expectations about intermediate forms and about what the fossil record could be expected to show. Our fossil collections contain what we have discovered of what *nature* has preserved. But:

Only a small portion of the surface of the globe has been geologically explored, and no part with sufficient care, as the important discoveries made every year in Europe prove. No organism wholly soft can be preserved. Shells and bones will decay and disappear when left on the bottom of the sea, where sediment is not accumulating.[17]

Moreover, although geological treatises foster an illusion that continuous deposition occurs in the same place over long stretches of time, closer scrutiny reveals that this is not so: "we know, for instance, from Sir R. Murchison's great work on Russia, what wide gaps there are in that country between the superimposed formations; so it is in North America, and in many other parts of the world."[18] When we think carefully about this phenomenon, we discover, Darwin claims, that the conditions under which fossiliferous beds will be formed are extremely demanding:

We may, I think, safely conclude that sediment must be accumulated in extremely thick, solid, or extensive masses, in order to withstand the incessant action of waves, when first upraised and during subsequent oscillations of level.[19]

Only two ways of forming such "masses" are known: either through deep sea accumulations (and few animals live in the depths) or through slow subsidence at the margins of shallow seas. Darwin concludes that "all our ancient formations, which are rich in fossils, have thus been formed during subsidence."[20] It follows that the geological record must be intermittent—as indeed Darwin's expert witnesses, Lyell and Forbes, have claimed. But there is a final twist. Periods of elevation of land enlarge the area of land "and of the adjoining shoal parts of the sea,"[21] thus providing opportunities for new varieties. Periods of subsidence, however, will be marked by extinctions with few opportunities for new forms. Hence, at the very times when fossils are being accumulated, novel species are less likely to arise. "Nature may almost be said to have guarded against the frequent discovery of her transitional or linking forms."[22]

The rhetorical strategy followed here is simple in conception, brilliant in execution. Darwin identifies an expectation that his readers will form about what character the history of life would have if his theory were true. He then poses a

series of considerations that undermine that expectation. The chapter culminates
with an extension of Lyell's famous image:

> For my part, following out Lyell's metaphor, I look at the natural
> geological record, as a history of the world imperfectly kept, and
> written in a changing dialect; of this history we possess the last volume
> alone, relating only to two or three countries. Of this volume, only
> here and there a short chapter has been preserved; and of each page,
> only here and there a few lines. [23]

If we reflect on the whole chapter in light of Darwin's metaphor, we can think of
the rhetorical strategy as beginning by presenting the reader with the idea of a
complete book. Darwin's successive arguments show that volumes must have
been discarded, chapters torn out, and lines obliterated.

I have concentrated so far on Darwin's efforts to persuade his readers that
descent with modification is possible. The third and most important task that he
must discharge is to make compelling the view that appeals to descent with
modification provide the best, even the only, explanations for many facts of
natural history. Two different responses need to be blocked. Potential readers
might recall from declarative memory some statements that would seem to provide
alternative explanations of the phenomena, or they might conclude that their
failure to do so signalled their own limitations of knowledge or imagination. The
first can be met by formulating explicitly the anticipated rival accounts and
showing their limitations. Here is a typical passage:

> It is difficult to imagine conditions of life more similar than deep
> limestone caverns under a nearly similar climate; so that on the
> common view of the blind animals having been separately created for
> the American and European caverns, close similarity in their
> organisation and affinities might have been expected; but, as Schiödte
> and others have remarked, this is not the case, and the cave-insects of
> the two continents are not more closely allied than might have been
> anticipated from the general resemblance of the other inhabitants of
> North America and Europe. [24]

Here there is a clear contrast between two rival ways of explaining facts about
distribution of animals—if the context is between special creation and descent
with modification, then the latter wins, for it can accommodate the details of what
we find in biogeography. Darwin reinforces the general point of difference
between the North American and European cave-insects with a telling detail:

> On my view we must suppose that American animals, having ordinary
> powers of vision, slowly migrated by successive generations from the
> outer world into the deeper and deeper recesses of the Kentucky caves,

as did European animals into the caves of Europe. We have some evidence of this gradation of habit; for, as Schiödte remarks, "animals not far remote from ordinary forms, prepare the transition from light to darkness. Next follow those that are constructed for twilight; and, last of all, those destined for total darkness."[25]

But how to address the worry that this is all very well as a refutation of creationist accounts, even the sophisticated ones popular among Darwin's contemporaries, but hardly a positive argument for descent with modification? Darwin pursues a two-fold strategy. His "intellectual pemmican" provides the reader with so much detail, so many instances in which descent with modification can account for the facts of comparative anatomy, biogeography, embryology, and so forth, as to force the conclusion that this is a powerful explanatory device. The second tactic is to circumscribe the space of possible explanations by emphasizing that among known processes only descent could possibly accommodate the phenomena. Consider the close of the description of South American animals:

> We ascend the lofty peaks of the Cordillera and we find an alpine species of bizcacha; we look to the waters, and we do not find the beaver or musk-rat, but the coypu and capybara, rodents of the American type. Innumerable other instances could be given. If we look to the islands off the American shore, however much they may differ in geological structure, the inhabitants, though they may be all peculiar species, are essentially American. We may look back to past ages, as shown in the last chapter, and we find American types then prevalent on the American continent and in the American seas. We see in these facts some deep organic bond, prevailing throughout space and time, over the same areas of land and water, and independent of their physical conditions. The naturalist must feel very little curiosity, who is not led to inquire what this bond is.
>
> This bond, on my theory, is simply inheritance, that cause which alone, as far as we positively know, produces organisms quite like, or as we see in the case of varieties nearly like each other. [26]

The effect of the catalogue of affinities among South American mammals, only the close of which I have quoted here, is to highlight the seriousness of the question of understanding them. Only the incurious will dismiss that question, for the "deep organic bond" has been made vivid through Darwin's descriptions. But to those who urge caution, who think that we must remain agnostic about what has produced this "deep organic bond," Darwin suggests that we know of only one possible process. The only alternative to accepting descent with modification is to speculate about some completely unknown process.

I have examined Darwin's solutions to three of the five rhetorical problems that he faced. Consideration of the other two leads into issues that I have so far slighted.

IV. Emergent Argument

Problem 2 recognizes that potential readers of the *Origin* might misunderstand the structure of Darwin's argument. Indeed, the subsequent responses to the effect that "Mr. Darwin has failed to produce the necessary proofs" and comments about "the law of higgledy-piggledy" testify that his critics formulated demands on an argument for descent with modification that they viewed the *Origin* as failing to satisfy.[27] After some years of debate, Darwin was able to find persuasive ways of formulating his argument, so that at least among British naturalists these failures of communication were overcome.

However, we should scrutinize the assumption that the argument that emerges at the close of the controversy about descent with modification is the argument that moved Darwin as he wrote the *Origin*. Inspired by Cordelia's vision of rhetoric, we are inclined to think that there was some cognitive process P that Darwin underwent and that led him to his conclusions about the relatedness of living organisms. His rhetorical task was to produce in the minds of his readers processes of the same type. Initially, he failed because his words prompted processes with different structures. Through more careful formulations, he was able to correct these unfortunate effects, so that his readers underwent processes of the same type as the process P that had initially moved him.

I want to suggest that the processes underlying belief in descent with modification themselves evolve during the debate of the 1860s. Moreover, these processes *improved* through Darwin's interactions with his allies and critics. By explicitly confronting various types of objections and by identifying the legitimate and illegitimate demands on his enterprise, there became available to Darwin and his fellow naturalists a process $P*$ of greater reliability than the process P that initially moved Darwin to his conclusions. Because of the commentary on it, the *Origin* was read differently in 1867 from the way in which it had been read in 1859, and Darwin's own reading was different. The "argument of the *Origin*" is thus an artifact of exchanges in which rhetorical devices play an important part.

Darwin's correspondence in the period following the publication of the *Origin* reveals his concern to understand the points on which his readers set greatest stock and the passages that caused them the most difficulty. His closest friends and allies offer advice, by describing what compels them. Thus Darwin's elder brother, Erasmus:

> To me the geographical distribution, I mean the relation of islands to
> continents is the most convincing of the proofs, and the relation of the

oldest forms to the existing species. I dare say I don't feel enough the absence of varieties, but then I don't in the least know if everything now living were fossilized whether the paleontologists could distinguish them.[28]

By contrast, those who are unconvinced provide Darwin with clues about how his case must be reworked. A recurrent charge is that he has speculated too far. Adam Sedgwick, Darwin's one-time geological mentor, wrote: "You have *deserted*—after a start in that tram-road of all solid physical truth—the true method of induction, and started us in machinery as wild, I think, as Bishop Wilkins's locomotive that was to sail with us to the moon."[29] This generic complaint was made in print in Hopkins' review, and it clearly prompted Darwin to rethink and to reformulate his conception of the requirements that the hypothesis of descent with modification must meet. After early complaints that Hopkins sets impossible standards for science[30] and that Hopkins has not understood him completely,[31] he offers a more precise diagnosis some six weeks later:

Hopkins's review in 'Fraser' is thought the best that has appeared against us. I believe that Hopkins is so much opposed because his course of study has never led him to reflect on such subjects as geographical distribution, classification, homologies, &c., so that he does not feel it a relief to have some kind of explanation.[32]

By the end of 1860, Darwin confessed to Huxley that he was "fairly sick of hostile reviews," but admitted that "they have been of use in showing me when to expatiate a little and to introduce a few new discussions."[33] He was now inclined to stress the power of his ideas to offer explanations of the facts of natural history, and to reiterate in these terms a call to the younger generation of naturalists that had been made in the first edition of the *Origin*:

I can pretty plainly see that, if my view is ever to be generally adopted, it will be by young men growing up and replacing the old workers, and then young ones finding that they can group facts and search out new lines of investigation better on the notion of descent than on that of creation.[34]

The emphasis on the unifying power of descent with modification recurs in Darwin's letters on reactions to the *Origin*. So, commenting to Hooker on Hutton's review, Darwin writes:

He is one of the very few who see that the change of species cannot be directly proved and that the doctrine must sink or swim according as it groups and explains phenomena. It is really curious how few judge it in this way, which is clearly the right way.[35]

Darwin's reflections on the form of his reasoning were ultimately expressed in an analogy with the wave theory of light, intended to make clear the standards that descent with modification was to satisfy and to respond to the charge that in setting such weak standards he was deserting the true path of science.

Quite independently of the particular "new facts" that could be introduced into the debate of the 1860s, the cognitive processes of Darwin and other evolutionists improved during this period as they came to recognize explicitly the value of "grouping and explaining the phenomena." Moreover, this was only one among many respects in which the interactions with allies and critics strengthened the argument for descent. By 1867 the improved reasoning had become accessible to the majority of naturalists in Britain and to a rising number in North America. New formulations would still be needed for audiences with different background associations—such as the French who read Darwin as recapitulating Lamarckian ideas of transformation.[36] Still others have been necessary in our day, to respond to the clamor of "Scientific Creationists." Reflections on the power and limits of particular rhetorical strategies continue to drive the refinement of the cognitive processes that generate belief in descent with modification.

V. Inflaming the Passions

Effective persuasion requires, as we have seen, a sense of the cognitive backgrounds of our targets and a concomitant anticipation of the responses that they are likely to make. But my focus so far has been on what they believe rather than on what they desire. I now want to consider ways in which attention to the emotional states of targets, their goals and desires, can be both necessary and productive in the epistemic business of science.

Let us start with the remaining rhetorical problem that I posed for Darwin, the problem of avoiding instant dismissal. The reality of this problem is apparent from Fitzroy's demented display at the British Association meeting. What happened when Fitzroy tried to read the *Origin*? It is easy to imagine that his goal of defending the truths of religion became activated and that perceived tension between this goal and Darwin's claims became so strong that he was literally unable to read. Others, no less devout than Fitzroy, were unprovoked, able to evaluate Darwin's claims more sympathetically, and their reactions testify to Darwin's rhetorical success.

Part of the task was to soothe potential passions. From the beginning, Darwin deliberately set his proposal within the context of a great question among naturalists, the "mystery of mysteries." All contemporary discussants would agree that the question of the origins of life could not be resolved by supposing the literal truth of the *Genesis* story. Where then was the additional conflict with religion in

supposing that species were related by descent rather than having been successively created? For Darwin's British readers, the vision of a deity who acts
indirectly through the institution of natural laws was thoroughly familiar, and there
would have been no felt need to invoke constant intervention in the created world.
Of course, the thesis of descent had implications for the status of our own species,
but Darwin deliberately underplayed these. The solitary reference to the origin of
Homo sapiens occurs two pages from the end, in Darwin's agenda for future
science: "Light will be thrown on the origin of man and his history."[37]

So one part of the solution is to avoid any appearance of conflict with the
doctrines of religion. Another is to deploy the preferred phraseology of religious
texts and of nineteenth-century natural theology to advance doctrines that might
otherwise seem threatening. Claiming the relatedness of all organisms, Darwin
consciously recapitulates Biblical phraseology: "Therefore I should infer from
analogy that probably all the organic beings which have ever lived on this earth
have descended from one primordial form, into which life was first breathed."[38]
Four pages later, he endorses Leibniz' view that God would have known better
than to have constantly to tinker with his creation, presenting it in the idiom of
nineteenth-century natural theologians:

> To my mind it accords better with what we know of the laws im-
> pressed on matter by the Creator, that the production and extinction of
> the past and present inhabitants of the world should have been due to
> secondary causes, like those determining the birth and death of an
> individual.[39]

The final sentence of the book combines this conception of a nobler creation with
a reprise of the suggestive image of the breath of primordial life: "There is
grandeur in this view of life, with its several powers, having been originally
breathed into a few forms or into one . . . ."[40]

Here, I think, we can see Darwin as a master of the "glib and oily art," and
legitimately accuse him of "speaking and purposing not." Given his own testimony about his religious beliefs, it is hard to credit his believing that creation
through secondary causes redounds especially to the nobility of the creator.
Darwin's correspondence reveals him as recognizing the controversial character
of the claim that there was one "primordial form of life," but the use of the Biblical
reference to "breath" is gratuitous and undiscussed.

Assuming that Darwin's tactics of appeasement rendered reactions like
Fitzroy's less frequent, we can ask if such tactics promote the human epistemic
project. The answer seems relatively obvious. If the presence of particular goals
can interfere with the epistemic evaluation of a novel proposal, then it is
epistemically desirable for the proposer to respond to those goals, even if it

requires deception. But now we should pose the converse question. Is it ever scientifically productive for a source to inflame the passions of potential targets, to present a proposal in ways that activate nonepistemic goals and desires?

Relatively mild examples of this rhetorical strategy can be found in Darwin's appeal to the rising generation of naturalists, in his carefully crafted praise of his friends, and in the inspiring vision of the future of biology that he presents towards the end of the *Origin*. "When the views entertained in this volume on the origin of species, or when analogous views are generally admitted, we can dimly foresee that there will be a considerable revolution in natural history."[41] The invitation to naturalists to join in an exciting and productive venture is reiterated throughout the following paragraphs: "the more general departments of natural history will rise greatly in interest";[42] "a grand and almost untrodden field of inquiry will be opened, on the causes and laws of variation, . . . ";[43] and so forth. As I read Darwin, these passages exhort aspiring naturalists to participate in his venture. "Join me," he seems to say, "and leave your mark upon science."

Such urgings are by no means unique. In parallel fashion Leibniz encouraged his followers to continue using the new calculus in solving geometrical, kinematic, and algebraic problems, rather than worrying about the difficulties of its foundations (see his letter to l'Hopital urging him to extend the calculus "because of the application one can make of it to the operations of nature, which uses the infinite in everything it does").[44] In our own century, major new approaches have been advertised as opening up vast new vistas for the scientific workforce (witness Wilson's manifesto for sociobiology). Indeed, the device is so familiar that it even works in a litotic variant: "It has not escaped our notice that the specific pairing we have postulated immediately suggests a possible copying mechanism for the genetic material."[45]

The idea that there are new worlds to conquer, especially for the young who have the daring not to be bound by tradition, makes an emotional appeal that can easily win support for fledgling programs of research. Suppose that the actual track records of these programs would not warrant preferring them to more established approaches. Then, it appears, rhetorical devices have served their grubby purposes, deceiving some scientists into overestimating the promise of a program. However, even if emotional persuasion shortchanges the targets, there may be valuable consequences for the community. Keeping minority viewpoints alive is often important to community-wide progress.[46] Where inflaming the passions protects ideas from being abandoned before they have had a chance to develop, the cognitive aims of the community may thus be promoted.

But what of the converse situation, in which a theory or research program is denounced because it is felt by some to have unwelcome consequences? Surely that is a use of rhetoric that interferes with our cognitive ambitions? Not necessarily. Consider one of the most strident attacks launched in the recent history of science, the denunciation of E. O. Wilson's *Sociobiology* by the Sociobiology Study Group of Science for the People. In their original letter to the *New York Review of Books*, members of the group began by linking Wilson to a tradition of "biological determinism" embracing Herbert Spencer, Konrad Lorenz, and Robert Ardrey.[47] Tracing the continued popularity of this tradition to its ability to "provide a genetic justification of the *status quo* and of existing privileges for certain groups according to class, race or sex," the letter recalled John D. Rockefeller's defense of unfettered capitalism through the invocation of Darwin and continued with an even more dramatic example:

> These theories provided an important basis for the enactment of sterilization laws and restrictive immigration laws by the United States between 1910 and 1930 and also for the eugenics policies which led to the establishment of gas chambers in Nazi Germany.[48]

Reverting to the same theme at the end of their letter, the authors charged:

> What Wilson's book illustrates to us is the enormous difficulty in separating out not only the effects of environment (e.g., cultural transmission) but also the personal and social class prejudices of the researcher. Wilson joins the long parade of biological determinists whose work has served to buttress the institutions of their society by exonerating them from responsibility for social problems.[49]

Quite understandably, Wilson was hurt by what he viewed as "ugly, irresponsible, and totally false accusation[s]."[50] Apart from the imputation to him of motivations that he denied, he feared that appraisal of a serious scientific research project would be distorted by appeals to political sympathies. Many other scientists concurred in the judgment that the credentials of sociobiology were to be assessed by considering the evidence, not its potential political consequences. So was the inflaming of liberal passions inappropriate?

Not entirely. We should distinguish two effects that the rhetoric of *Science for the People* had: (i) rejection of sociobiology out of hand; (ii) closer scrutiny and debate about the evidence for sociobiological claims, particularly as they apply to human beings. There are probably some people who, like Fitzroy with respect to the *Origin*, dismissed sociobiology because of the political associations emphasized by *Science for the People*. But the dominant response among scientists, social scientists, and philosophers was to probe more carefully the reasoning that Wilson and some of his followers offered. Precisely because the potential human consequences of widespread acceptance of Wilson's conclusions were so significant, close examination of the arguments was valuable.[51] In the past decade, evolutionary studies of social behavior in general and of human social behavior in particular have been greatly refined because of the interchanges among defenders and critics of human sociobiology. Our improved perspective on the relevant issues is, I think, traceable to the debate that *Science for the People* sparked and to the passions that its rhetoric inflamed.

VI. Dialogues of One

My aim so far has been to delineate the uses of rhetoric in the construction of scientific arguments through social exchanges. But as I noted earlier, I believe that there are similar uses in scientists' individual attempts to work out their own views. Persuasion does not only go on between scientists but between stages of the same scientist at different times.

For those working out complex and novel views, ways have to be found to summarize parts of the overall argument. Actual thought processes are surely more intricate than the records that scientists leave in their notebooks, but even these reveal the uses of striking images and the rehearsal of rhetorical tactics. Consider Darwin's early notebook on "transmutation of species," Notebook B, dating from 1837 to 1838.[52] Many of the entries summarize a view, a conclusion, a line of argument, or an image, which then led to specific questions that Darwin subsequently pursued.

Early in the notebook, we find a succinct version of the thesis that the idea of descent with modification can account for facts of island biogeography:

> According to this view animals, on separate islands, ought to become different if kept long enough.—"apart, with slightly differen circumstances.—" Now Galapagos Tortoises, Mocking Birds; Falkland Fox—Chiloe, fox,—Inglish & Irish Hare.—[53]

Darwin's formulation of a possible source of evidence for transmutationist views is accompanied by a list of possible exemplars. Considering further how best to illustrate and develop the point, a slightly later entry continues:

> . . . island near continents might have some species same as nearest land, which were later arrivals/others old ones, (of which none of same kind had in interval arrived) might have grown altered
>
> Hence the type would be of the continent though species all different. In cases as Galapagos & Juan Fernandez. When continet of Pacific existed might have been Monsoons . . . when they ceased importation ceased &/changes commenced.—or intermediate land existed.—or they may represent some large country long since separated.—[54]

So Darwin begins to pose questions about connection of land masses and about means of transportation. Some pages later, he introduces different examples.

> I should expect that Bear & Foxes &c same in N. America & Asia, but many species closely allied but different, because country separated since time of extinct quadrupeds:—same argument applies to England.—Mem. Shew Mice.—[55]
>
> Geographic distribution of Mammalia more valuable than any other, because less easily transported—Mem plants on Coral islets.— Next to animals land birds.—& life shorter or change greater—
>
> In the East Indian Archipelago it would be interesting to trace limits of large animals—
>
> Owls transport mice alive?[56]

As I read these disjointed comments, Darwin is engaged in a process of self-persuasion that depends crucially on descriptions of phenomena that will focus his subsequent inquiries. *It is simply not obvious in advance of these remarks what will constitute evidence for or against the thesis of descent with modification.* Darwin has a sense that his observations on the *Beagle* voyage and some of the observations of other naturalists bear on the issue. The notebooks show him trying to develop this sense, to articulate his half-formed thoughts, so that they emerge as an argument Just as the full argument for descent with modification was the end product of persuasive exchanges in the wake of publication of the *Origin*, so the reasoning that Darwin developed in the 1840s and 1850s was the product of persuasive conversations among his earlier selves, conversations that were crucially facilitated by the notebook jottings. Those notebooks effectively enable us to eavesdrop on a "dialogue of one."

VII. Conclusions

I shall close by summarizing the position that I have been trying to explore. Traditional thinking about logic and rhetoric opposes the forms of correct thought to the verbal dress in which they appear. That conception would be appropriate for a certain type of cognitive system. If we were equipped with inferential propensities that would enable us to draw the right conclusions from evidence no matter how presented, then the language in which phenomena were reported and arguments made would play no essential role, and our concern with rhetoric in science would be a purely negative one. We should be on guard against the "glib and oily art" that might interfere with the deliverances of reason. However, for limited cognitive systems, such as ourselves, activating the right inferential propensities and thus reaching the correct conclusions always depend on the mode of presentation of the scientific material. In some instances this is entirely unproblematic. Conventionalized forms of rhetoric—"dead rhetoric"—serve us admirably, and no creative effort is required. But those, like Darwin, who venture into uncharted territory have to provide ways of seeing what they take to be relevant. With new images, metaphors, and well-designed forms of argument, they have to persuade us to see just what bears on their conclusions. Through persuasive exchanges, crystallized forms of "the argument" finally emerge.

We, like Cordelia, have been beguiled by a contrast between transparent and disguised presentations of thought, between proof and deception. We have over-looked the omnipresence of persuasion, conceived as the art of inducing reliable belief-forming processes. If we are forced to abandon traditional ideas about "rigorous scientific thought," we may console ourselves by realizing that persuasion is nowhere near as bad as its traditional reputation might suggest. Indeed, as the opening scene of *King Lear* attests, "speaking plainly" has its own (perhaps unintended) persuasive effects. Cordelia was wrong. Rhetoric is inescapable. [57]

NOTES

1. William Shakespeare, *King Lear*, I–i–223–232.

2. Perhaps in doing so, I return the idea of rhetoric to a yet older tradition, that stemming from Aristotle. Limitations of competence prevent me from being able to evaluate the exact degree of kinship between Aristotle's claims about rhetoric and argument and those I offer below. However, Marcello Pera's illuminating discussion in his contribution to this volume encourages me to think that there is some affinity to be traced.

3. Here I offer without defense a standard for assessing the merits of cognitive processes that is familiar within general epistemology. See, for example, Alvin Goldman, *Epistemology and Cognition* (Cambridge, Mass.: Harvard University Press, 1986). More exact formulations of my preferred approach are found in my article "The Naturalists Return" (to appear in *The Philosophical Review*, 1992), and in a book currently in preparation. For the present, this simple version of reliabilism will do.

   However, it should be noted that my advocacy of reliabilism marks the point at which I diverge from Marcello Pera's insightful treatment of rhetoric and persuasion (see the later sections of his contribution to this volume). Where he proposes to understand the goodness of arguments in terms of their social power, I hearken back to the idea that some forms of reasoning are more likely to promote true belief (more generally: epistemically virtuous beliefs) than others. As Dudley Shapere noted, this means that for all their liberality towards rhetoric in science my proposals retain the old-fashioned idea of scientific objectivity. I believe that this eclectic mix is possible once one appreciates the importance of doing epistemology within a psychologistic framework. See my "The Naturalists Return," for more extended discussion.

4. I would prefer to think of skill memory as consisting of nested systems of production rules, in the manner explained by John Anderson in *The Architecture of Cognition* (Cambridge, Mass.: Harvard University Press, 1983). Such systems capture many of the features of Marvin Minsky's "frames" or Roger Schank's "scripts." As Peter Machamer forcefully pointed out in his comments on my presentation at this conference, many of the claims I make about the reception of Darwin's *Origin* can be amplified by thinking explicitly in terms of activation of components of a frame or of parts of a script. For reasons of space, I shall not offer such explicit treatment here, but simply register my agreement about the possibility and the desirability of doing so.

5. G. H. Hardy, *A Mathematician's Apology* (Cambridge: Cambridge University Press, 1941).

6. I chose the term "dead rhetoric" by analogy with "dead metaphor" to suggest that frequency of usage can make the rhetorical function of a device invisible. At this conference, R. S. Westfall pointed out to me that 'dead' has unfortunate connotations and suggested that I use some such term as 'established,' 'traditional,' or 'fixed.' I agree with him that my terminology is imperfect, but have chosen to stick with it because of its indications of the invisibility of some pieces of scientific rhetoric. Once that point is firmly appreciated, "established rhetoric" will do far better.

7. Steven Shapin, "Pump and Circumstance: Robert Boyle's Literary Technology," *Social Studies of Science*, 14 (1984), pp. 481–520.

8. For discussion of another example in a fashion complementary to that pursued in the text, see Robert Westman, "Proof, Poetics, and Patronage: Copernicus' Preface to *De Revolutionibus*," in *Reappraisals of the Scientific Revolution* David Lindberg and Robert Westman, eds. (Cambridge: Cambridge University Press, 1990).

9. Charles Darwin, *On the Origin of Species by means of Natural Selection, or the Preservation of Favoured Races in the Struggle for Life* (London: John Murray, 1859). My references will be to the facsimile of the first edition with introduction by Ernst Mayr (Cambridge, Mass.: Harvard University Press, 1964).

10. I have defended this interpretation of Darwin's argumentative strategy in "Darwin's Achievement," in *Reason and Rationality in Science*, Nicholas Rescher, ed. (Washington, D.C.: University Press of America, 1985), pp. 123–185. My reconstruction accords with that sketched by T. H. Huxley in *Darwiniana* (New York: Appleton, 1896), p. 72, and articulated by M.J.S. Hodge in "The Structure and Strategy of Darwin's 'Long Argument'," *British Journal for the History of Science*, 10 (1977), pp. 237–246.

11. See Hopkins' review of the *Origin*, reprinted in *Darwin and his Critics*, David Hull, ed. (Cambridge, Mass.: Harvard University Press,   ), p. 249.

12. Huxley, *Darwiniana*, p. 25.

13. While the *Istituto Filosofico* was providing excellent hospitality to those of us discussing science and rhetoric, Italy was showing the world how to host a major international sporting event, one that puts into proper perspective American rhetori-

cal appropriations of such names as "World Series" and "Superbowl." I have retained the topical example.

14. *Origin*, p. 84.

15. *Ibid.*, p. 280.

16. *Ibid.*

17. *Ibid.*, p. 288.

18. *Ibid.*, p. 289.

19. *Ibid.*, p. 290.

20. *Ibid.*, p. 291.

21. *Ibid.*, p. 292.

22. *Ibid.*

23. *Ibid.*, pp. 310–311.

24. *Ibid.*, p. 138.

25. *Ibid.*

26. *Ibid.*, pp. 349–350.

27. Complaints about the failures of Darwin's "proofs" run throughout the responses of William Hopkins, Richard Owen, and Adam Sedgwick, all reprinted in Hull, *Darwin and his Critics*. John Herschel delivered the comment about the "law of higgledy-piggledy."

28. *The Life and Letters of Charles Darwin*, three volumes, Francis Darwin, ed. (London: John Murray, 1888; New York: Johnson Reprint Co., 1969), vol. 2, p. 233.

29. *Ibid.*, p. 248.

30. *Ibid.*, p. 315.

31. *Ibid.*, pp. 317–318.

32. *Ibid.*, p. 327.

33. *Ibid.*, p. 354.

34. *Ibid.*, p. 355.

35. *Ibid.*, p. 362.

36. See the essays in *The Comparative Reception of Darwinism*, Thomas Glick, ed. (Austin: University of Texas, 1974) and the contributions by Peter Bowler, Pietro Corsi and Paul Weindling, and Francesco Scudo and Michele Acanfora to *The Darwinian Heritage*, David Kohn, ed. (Princeton: Princeton University Press, 1986). In the discussion in Ischia, Gerald Holton rightly emphasized the importance of relativization to audiences (see his contribution to this volume). The differences in the reception of Darwin's argument among different groups provide further grist for Holton's mill.

37. *Origin*, p. 488.

38. *Ibid.*, p. 484.

39. *Ibid.*, p. 488.

40. *Ibid.*, p. 490.

41. *Ibid.*, p. 484.

42. *Ibid.*, p. 485.

43. *Ibid.*, p. 486.

44. G. Leibniz, *Mathematische Schriften*, C. Gerhardt, ed. (Halle, 1849–1863), vol. 2, p. 219.

45. J. Watson and F.H.C. Crick, "A Molecular Structure for Deoxyribonucleic Acid," reproduced in *The Double Helix*, Gunther Stent, ed. (New York: Norton, 1978), p. 240.

46. I argue this point in detail in "The Division of Cognitive Labor," *Journal of Philosophy*, 87 (1990), pp. 5–22.

47. This letter is reproduced in *The Sociobiology Debate*, Arthur Caplan, ed. (New York: Harper and Row, 1978), pp. 259–264. See especially pp. 260–261.

48. *Ibid.*, p. 260.

49. *Ibid.*, p. 264.

50. See E.O. Wilson, "For Sociobiology," in *ibid.*, pp. 265–268, 265.

51. See the discussion of this point in my *Vaulting Ambition: Sociobiology and the Quest for Human Nature* (Cambridge, Mass.: MIT Press, 1985), pp. 8–10.

52. Charles Darwin, *Notebooks*, Sydney Smith, ed. (Cambridge: Cambridge University Press, 1988).

53. *Ibid.*, p. 7.

54. *Ibid.*, pp. 10–12.

55. *Ibid.*, p. 65.

56. *Ibid.*, pp. 81–82.

57. I am grateful to those who participated in a lively discussion of the oral presentation of this paper at Ischia, for their comments and advice, not all of which I have been able to follow. Thanks are especially due to Gerald Holton, Peter Machamer, Marcello Pera, Dudley Shapere, and R. S. Westfall, all of whom made valuable suggestions.

The Role and
Value of Rhetoric
in Science

MARCELLO PERA

I. Back to Aristotle

The historical fact that Aristotelian science has been defeated by Galileian science and "Aristotelian method" replaced by "Galileian method" (whatever these two philosophical and highly rhetorical figures may look like) does not mean that Aristotle has no longer anything to teach us as regards the way science proceeds. In this paper I shall try to go back to some of his ideas and to use a revised version of them as a contribution to the debate on the so-called "crisis of method" in the recent philosophy of science. In one sense, my tribute to Aristotle is a pretext, for what I have to say does not stem directly from his view and differs from it in several respects; in another sense, it is a genuine homage because I believe that Aristotle lets us envisage an image of scientific inquiry that, properly adapted, may be taken as an efficacious antidote against the two opposite diseases that currently affect the philosophy of science, namely, methodological dogmatism and socio-logical or postphilosophical mannerism.

Aristotle's standard, textbook view is that science is based upon demon-stration (*apodeixis*). This view is not incorrect, but it is only part of the story. For Aristotle also maintained that dialectics and rhetoric play a role in scientific knowledge. According to him, both these techniques and ways of arguing have a cognitive function. Not only is dialectics useful "in relation to the ultimate bases of the principles used in the several sciences," for "it is through the opinions generally held on the particular points that these have to be discussed"[1]; dialectics is also an integral part of scientific inquiry because "the ability to raise searching difficulties on both sides of a subject will make us detect more easily the truth and error about the several points that arise."[2] In the same way, rhetoric shows us how to "employ persuasion, just as strict reasoning can be employed, on opposite sides of a question, not in order that we may in practice employ it in both ways (for we must not make people believe what is wrong), but in order that we may see clearly what the facts are, and that, if another man argues unfairly, we on our part may be able to confute him."[3]

Aristotle established the distinction between dialectics and rhetoric in many ways.[4] The most general is that the former concerns the capacities or abilities (*dynameis*) of rightly using arguments (*syllogismoi*), the latter the capacities or abilities of properly using the ways of persuasion (*pisteis*). But since the ways of persuasion—especially the technical ones, which consist of arguments—are "a sort of demonstration" (*apodeixis tis*),[5] rhetoric is an image (*homoioma*) of dialectics or, as Aristotle also says, a counterpart (*antistrophos*)[6] of it.

More important is the distinction between dialectics-rhetoric and demonstration. In demonstrative syllogism, we start from premises that are true and universal, while in dialectical and rhetorical argument, we proceed from premises that are widely admitted and shared opinions (*endoxa*). Demonstration draws true conclusions directly from definitions that grasp essences; dialectical and rhetorical argumentation draws conclusions from accepted opinions and discussion of rival theses. Therefore, the procedure of science, in particular of empirical sciences, that is, those sciences that "start from the essence" by "making it plain to the senses" and not by "assuming it as a hypothesis,"[7] is a sequence of this kind:

A. sensation—definition—demonstration—knowledge.

The third link connecting the steps of this sequence is equality (the conclusion of a scientific syllogism *is* knowledge); the second is syllogism ("by demonstration I mean a syllogism productive of scientific knowledge")[8]; and the first is induction. As Aristotle maintains in a famous passage, since "the basic premises of demonstrations are definitions"[9] that cannot be known through demonstration, "it is clear that we must get to know the primary premises by induction; for the method by which even sense-perception implants the universal is inductive."[10]

But if dialectics, too, has a cognitive value and plays a role in science, it is to be expected that in addition to procedure A science also makes use of another. This, we might presume, will happen whenever a phenomenon does not universally or patently make its corresponding essence plain, and therefore a problematic situation (*aporia*) arises. In such cases, the procedure for arriving at knowledge is a sequence of this kind:

B. sensation—problem—dialectics—knowledge.

Unfortunately, here, as in many other cases, Aristotle leaves his interpreters in the dark as regards the function of this procedure. We seem to have two options. The former is that B is subservient to A, in the sense that the segment of sequence "problem—dialectics" in B replaces or encompasses the inductive link in A. This, as we have seen, is in perfect agreement with what Aristotle says in the *Topica* about how the principles of science can be attained, although it is different from what he rather dogmatically states in the *Analytica posteriora* about the same topic. We may combine the two passages by taking induction in A in a broad sense, as a kind of transition from sensation to the principles, not necessarily of the enumerative form, which is a sensible, reasonable reading. But there is another option that also has some textual evidence. We may take B as a sequence *per se*, one that is suitable to those kinds of domains in which it is difficult to grasp an essence or where there are different opinions about the proper essence. This option is supported by Aristotle's frequent recourse to procedure B in his writings on physics and especially cosmology, [11] which seem to be domains of such a kind.

Consider, for example, Aristotle's proofs of the incorruptibility and ingenerability of the universe. He says that we have to "start with a review of the theories of other thinkers; for the proofs of a theory are difficulties for the contrary theory. Besides, those who have first heard the pleas of our adversaries will be more likely to credit the assertions which we are going to make." [12] Or, consider how he proves the immobility of the Earth. [13] Aristotle starts with a dispute with his predecessors, confuting five different theories; then he goes on to a dialogue with himself, considering the two contradictory theses that the Earth either stands still or it moves; finally he rejects the latter thesis. In these cases, procedure B does not seem subservient to A. Here, discussion and refutation lead straight to knowledge, not straight to essence and then indirectly to knowledge.

Whatever option we resolve upon, it is clear that procedure B does not banish the role of experience. Not only does Aristotle often resort to observation in order to reject a thesis, [14] he also stresses that confuting rival theses by raising "searching difficulties on both sides of a subject" is not the same as a mere verbal discussion of opinions at hand, for it involves a critical examination of all the theses and objections that are "proper" to the subject itself. [15]

That many Aristotelians and Scholastics during the Middle Ages weakened the role of experience in procedure B to such an extent that science was often reduced to verbal disputations of, or dogmatic oaths on, Aristotle's opinions is well known. It is also well known that modern science started with a reaction against (the degeneration of) this way of pursuing scientific inquiry. Bacon, for example, banned dialectics because, according to him, dialecticians are interested in dialogue and dispute and not in the concrete things dialogue and dispute are about. The aim of science, Bacon wrote, is "to overcome, not an adversary in argument, but nature in action";[16] only those sciences that are founded on opinions and dogmas can make use of dialectics, "for in them the object is to command assent to the proposition, not to master the thing."[17] Galileo had a similar opinion. He wrote that "in the natural sciences the art of oratory is ineffective" because science deals with conclusions that are "true and necessary," and "a thousand Demosthenes and a thousand Aristotles would be left in the lurch by every mediocre wit who happened to hit upon the truth for himself."[18] Therefore, we should "not entangle flowers of rhetoric in the rigors of demonstration."[19] The idea Bacon insisted upon is that if we want to obtain genuine knowledge, we must hold a dialogue with nature, not with philosophers, that is, we must observe and carry on experiments and not disputes. In this dialogue, the opinions of philosophers are not necessary; rather, those are idols to be demolished. For his part, the idea Galileo insisted upon is that the language with which nature has to be questioned is mathematics, for mathematics is the very language nature is written in.

By replacing syllogism with mathematical proofs, Galileo and his followers transformed Aristotle's procedure A into:

C. observation—axiom—deduction—knowledge.

According to Galileo, this procedure holds for applied mathematics, especially mechanics, as he shows in his *Notes* upon Antonio Rocco's *Esercitazioni filosofiche*, where he clearly describes the four steps through which he (maintained to have) arrived at the law of falling bodies.[20]

By replacing dialectical dispute with mathematical inquiry, Galileo transformed Aristotle's procedure B into:

D. observation—hypothesis—deduction—knowledge.[21]

Galileo maintained that this procedure holds good for physics and especially for cosmology,[22] where axioms cannot be reached and must be replaced by tentative hypotheses to be submitted to the test of experience.

But advancing hypotheses and deducing consequences from them is not always enough to attain knowledge. Galileo himself had to face the problem of two rival theories leading to the same observational consequences. In these situations, procedure D is clearly too weak, and other factors have to be resorted to. This next step was accomplished by the subsequent philosophical reflection on science that transformed D into the following sequence, usually referred to as the "hypothetico-deductive method":

E. observation—hypothesis—method—knowledge.

Here the first link is taken sometimes as logical (inductive, abductive) and sometimes as alogical (creativity, free invention), while, depending on whether the distinction between "context of discovery" and "context of justification" is made, the rules of method are understood either as norms of invention and the generation of hypotheses or, as in modern interpretations of this procedure, as norms of proof of the conclusion. In both cases, method is required to fill in the gap between experience and knowledge.[23]

It is important to note that while procedure C is equivalent to A, the only difference being that mathematics replaces syllogism, D and even more so E are not the same as B. Granted, method inherits from dialectics its function of settling disputes; it also inherits its juridical metaphors, for method is taken as judge in a tribunal, just like dialectics was taken as arbiter.[24] But method and dialectics are not the same because they belong to two different views about knowledge and rationality, which we may call the *dialogical view* and the *methodological view*, respectively. According to the former, knowledge is a game with two players: nature, which provides experience, and a single (or collective) inquiring subject who tries to decipher or shed light upon it;[25] according to the latter, there are three players: nature; the inquirer, who asks questions; and he (or those) who, by questioning nature, too, and disputing, decide(s) about the right or most acceptable answer. If this dispute is to be omitted because what counts is not "to overcome an adversary in argument but nature in action," experience alone is the ground of knowledge. But if experience alone is the ground of knowledge, dialectics is to be replaced by method, for method is precisely the judge of knowledge-claims grounded on experience and the technique of questioning it.

This technique may be understood differently. It may consist of a mathematical calculus, as in Galileo's "necessary demonstrations," or of a logical calculus, as in Leibniz' probabilistic "scales of proofs," or of a set of "certain and simple rules," as in the case of Descartes' "rules for the direction of the mind." In all these cases, method performs the same function: it lead us, all of us, to the knowledge of nature as it is, independently of our opinions. Method is *impersonal*. Bacon writes that his method "levels man's wits,"[26] like a rule or compass. Of his rules, Descartes maintains that "if a man observes them accurately, he shall never assume what is false as true."[27] In the same vein, Leibniz holds that, thanks to his scales of proof, "all truths can be discovered by anybody and with a secure method (*methodo certa*)."[28] In modern times, Popper and Lakatos take method as a set of rules that allows us to draw a "clear line of demarcation between science and metaphysical ideas"[29] or to obtain "sharp criteria"[30] for comparing rival theses. Thus, method is also *universal*. It depends neither on individuals nor on epochs; as Lakatos says, it provides us with "universal definitions of science."[31] Lastly, method is *a priori*, for it is prior to knowledge and independent of its growth. According to Descartes, the rules of method "contain the primary rudiments of human reason";[32] according to Galileo, mathematical knowledge is (intensively) divine knowledge.[33]

The differences between the dialogical view and the methodological view are relevant. Dialectics conceives of knowledge as the outcome of a concrete dispute between interlocutors holding rival theses; method as the outcome of an ideal, neutral confrontation of these theses with a single yardstick. Dialectics needs an audience with a framework of shared opinions; method gets rid of both the audience and the need for shared opinions. Therefore, dialectics proves in a less cogent way: its conclusions are reasonable or convincing; method proves in a more necessary way: its conclusions are infallibly valid or invalid, or probable to this or that degree. From the point of view of dialectics, scientific rationality is a historical, cultural property; from the point of view of method, it is ahistorical. The former looks at science from a human viewpoint, the latter through God's eyes.

Needless to say, these eyes proved to be worth using. But if we do not believe that the victory of method over dialectics is a necessary fact, then the present-day failure, or even bankruptcy, of method allows us to put new life into the old dialogical view. Of course, we cannot proceed as if much, important (methodological) water had not passed under our (scientific) bridges. If, as I suggest, we embark upon a dialectical or rhetorical project, this has to be a project for a *new* dialectics or rhetoric. But Aristotle can still show us the main steps to be taken.

The following elements were essential to the old dialectical view of science: 1. the means of conducting discourses (*dynameis tou porisai logous*), that is, the forms of persuasion (*pisteis*) and reasoning (*syllogismoi*) the interlocutors make use of in a dispute; 2. the accepted opinions (*endoxa*) and what is considered to be convincing (*pythanon*), that the interlocutors in a dispute refer to; 3. the refutation (*elenchos*) and opposition (*enantion*), that is, the ways and rules of conducting disputes and opposing and refuting the interlocutors; 4. the dialectical strength (*dialektike ischys*), that is, the force and value of dialectical conclusions.

Accordingly, a project for a new dialectics or rhetoric of science has to face the following problems: 1. to review the argumentative forms made use of in scientific disputes and the functions they perform; 2. to fix the substantive factors that the interlocutors of a dispute refer to in order to reach their conclusions; 3. to fix the procedural factors in terms of which disputes are conducted and settled; 4. to establish the force of scientific arguments and the source of the other evaluative notions we refer to with such expressions as "good," or "sound," or "strong" scientific argument.

My (ambitious) aim in this paper is to take all these steps, however quickly. I shall try to show why scientists resort to rhetoric and in what contexts. Then I shall endeavor to outline what I call the "basis of scientific rhetoric," that is, the factors in terms of which rhetorical arguments can be evaluated. My last concern will be the main advantages to be expected from this project. If I am not wrong, it allows us to envisage a new form of scientific rationality that is less rigid than the one recommended by the methodologists, less elastic than the one suggested by the counter-methodologists, but more suitable than both. If I am allowed to use some rhetoric, this may be enough to make the project attractive.

II. The Functions of Rhetoric in Science

According to Aristotle, there is no relevant, technical distinction between dialectics and rhetoric because both make use of the same kinds of arguments. This is a point where we must part company with him. Rhetorical arguments do not boil down to syllogisms and inductions alone, or enthymemes and paradigm arguments, which Aristotle considers as their counterparts in the realm of persuasive argumentation. There are many more ways of arguing with interlocutors in a dispute and persuading them. Aristotle himself, when he actually practices dialectics and not only theorizes, employs different kinds of arguments. For example, when he contrasts the view that the earth moves, he says that if this were true, then the earth would move with more than one movement, like all the celestial bodies, which is a typical analogical argument.[34]

Moreover, if we take the standard Aristotelian definitions of rhetoric and dialectics as the art of persuading and the art of arguing or confuting, respectively, we have to depart from Aristotle again. Rhetoric is not the counterpart of dialectics, rather it includes it. At least in science, persuading an audience or converting it includes contrasting rival opinions. For example, when a scientist tries to convince his interlocutors that a certain hypothesis H is a plausible explanation for a set of phenomena E, or that a theory T is a promising research project for a set of empirical and theoretical problems P, he can hope to attain his goal only if he compares H and T with competing explanations and solutions of E and P and is able to reject them.

In this paper, then, scientific rhetoric will be considered as the art of making use of persuasive arguments in order to change or reinforce opinions in a scientific community about questions that have cognitive value. A persuasive or rhetorical argument is an argument that is neither formally stringent nor empirically compelling. As I take it, then, scientific rhetoric is the set of those persuasive, argumentative techniques scientists use in order to reach their conclusions, not the modes of expression, or the ornament, or the style that may accompany those arguments.

That these techniques have been neglected for so long is due to the now standard classification of arguments in deductive and inductive and to an overestimation of the role of deduction and induction for cognitive purposes. But the standard classification is too narrow, for it leads us to consider as fallacies a lot of arguments that in relevant contexts are legitimate and accepted, and that perform relevant functions. Consider, for example, an argument of the form "X says p; X is not a reliable source; p is not worth crediting." From a formal point of view, this argument is invalid, but in many contexts it is taken as a good argument. It is the typical argument by which prosecutors try to convince the members of a jury that a certain witness should not be taken into consideration.

On the other hand, deduction and induction are clearly poor tools for serving all the purposes of scientific research. Consider the following situation: A scientist draws a conclusion O from a theory T; he carries out an experiment and obtains the result O, then he concludes that T is confirmed and acceptable. This conclusion presupposes some preliminary decisions. For example, the scientist has to decide whether the experiment is reliable, whether O is a severe test, whether empirical confirmation is the only desideratum T has to fulfil, and so on. With what kind of arguments will the scientist argue for these decisions? Obviously, he makes use of deduction when he infers O from T, and of induction when he concludes that, given O, T is probable. But these deductive and inductive inferences are of no use to the scientist for his preliminary decisions. Such inferences come *later*. The arguments he needs in order to justify these decisions are neither deductive nor inductive. As he will have to develop considerations of opportunity (for example, is it right to accept a hypothesis on the basis of its empirical confirmations alone?) and value (for example, are other epistemic desiderata for theory-acceptance as important as empirical confirmation or even more so?), such arguments will be typically rhetorical.

We encounter certain recurrent rhetorical arguments in science if we examine different contexts of scientific reasoning. The following are the most important.

Choosing a style or line of inquiry. This question arises especially when a new theory is associated with a new way of doing research. Here the burden for innovators is terribly heavy. They have to fight on two fronts, for they know that their own theory has no chance of being accepted if the new style is not agreed upon, but they also know that the main obstacle to the new style is precisely the theory it is linked with. Sometimes they will start with the theory and hope for something like a "drag effect"; at other times, they will engage in a "discourse on method." In this case, they will often make use of *ad hominem* arguments and arguments of retortion; that is, they will try to throw light upon a contradiction between what the critics *say* and what they actually *do* and to show that what they do is not only the opposite of what they say but precisely the application of the very method or style they oppose.

Let us consider an example. One line of attack against Darwin's theory of natural selection was based on methodological considerations. Many critics (rightly) objected that it was not proved according to strictly inductive Baconian standards. Even though later, in his *Autobiography*, Darwin wrote he had "worked on true Baconian principles,"[35] he was aware that his theory was defective from the point of view of Bacon's standards. But he had a good reply. In a passage added to the first edition of the *Origin*, he wrote:

> [1] It has recently been objected that this is an unsafe method of arguing; but it is a method used in judging of the common events of life, and has often been used by the greatest philosophers. The undulatory theory of light has thus been arrived at; and the belief in the revolution of the earth on its own axis was until lately supported by hardly any direct evidence. It is no valid objection that science as yet throws no light on the far higher problem of the essence or origin of life. Who can explain what is the essence of the attraction of gravity? No one objects to following out the results consequent on this unknown element of attraction; notwithstanding that Leibniz formerly accused Newton of introducing "occult qualities and miracles into philosophy."[36]

One of Darwin's critics, William Hopkins,[37] had observed that "he who appeals to Caesar must be judged by Caesar's laws." Darwin retorts that precisely Caesar's laws discharge him and damn his critics. And since the method they question is exactly the same as the one they practice, their objection is self-refuting or self-phagic.

Interpreting an admitted rule. The problem here is the exact interpretation of the prescriptive content of a rule of inquiry (or of the value from which such a rule stems). Take the rule of refutation (the underlying value being agreement with observations). Although two scientists may agree that if a hypothesis H has a logical consequence O_1 contradicted by empirical evidence O_e, then H is to be refuted, they may disagree about the relevant sense of "contradiction." For example, one may say that O_1 is contradicted by O_e if it is different from it, while another may maintain that O_1 is not contradicted if the difference between O_1 and O_e is very small and negligible. In these cases, the proponent of the hypothesis typically resorts to a pragmatical argument in order to persuade his interlocutors. He will try to weaken the rigor of the rule by appealing to certain future advantages, or by reminding his interlocutors that in the past, too, there have been situations in which the validity of the rule has been suspended with relevant gains or minor losses.

Consider how the cosmologist Herman Bondi, in the middle of the cosmological controversy in the 1950s, replies to the objection that his (and Gold's) steady-state theory of the universe is in contrast with the principle of conservation of energy, and therefore has to be rejected because it conflicts with the rule that theories are not to be accepted if they are in conflict with well-grounded, accepted laws. Bondi writes:

> [2] Dr. Bonnor has argued that this process of continual creation violates the principles of conservation of energy. . . . Now, in fact, the mean density in the universe is so low, and the time scale of the universe is so large, by comparison with terrestrial circumstances, that the process of continual creation required by the steady-state theory predicts the creation of only one hydrogen atom in a space the size of an ordinary living-room once every few million years. It is quite clear that this process, therefore, is in no way in conflict with the experiments on which the principle of the conservation of matter and energy is based.[38]

Applying a rule to concrete cases. A rule (or its corresponding value) may be accepted, its interpretation agreed upon, but the question may still arise whether it is pertinent to the case being discussed. Such questions may be difficult to solve, especially when one objects that the case is special and different from the one contemplated by the rule in the abstract. The person who raises these objections is in the same situation as a lawyer who argues that the deeds committed by his client do not fall within the bounds of such and such a law or that the application of such and such a law would be unfair.

Consider the following argument by William B. Bonnor, again from the cosmological debate:

> [3] Science would never voluntarily adopt hypotheses which restrict its scope. Sometimes restrictions are obligatory, as for example in the case of the Uncertainty Principle, which restricts the accuracy of certain physical measurements, but unless it is shown that such limitations apply to cosmology we should, I think, assume that our knowledge of the universe can stretch indefinitely into the past and into the future.[39]

What is under discussion here is the rule that restrictions should not be imposed on scientific hypotheses. As Bonnor supports a relativistic model of the universe, he rejects the perfect cosmological principle, which would impose a restriction on cosmology and lead to a different theory. Here he does not face the principle directly, but the rule. He admits that there are certain, special cases in which it is violated, but denies that cosmology should be one such case. His is an argument from ignorance. He argues that, unless one proves that cosmology is a special case, the general rule holds good for it, too. The burden of proof is on the opponents' shoulders. It is to them to prove that cosmology should be an exceptional case.

Justifying a starting point. To obtain a result we have to start from premises. The more generally accepted the premises, the more credible the result. But how can the premises be justified? As Aristotle maintained, dialectics serves this purpose well. If a premise is questionable, one will try to convince his interlocutors to admit it at least tentatively, hoping that the results will render it more acceptable later. Often scientists argue on the basis of the saying that "pains of childbirth are soon forgotten." In this case, it may turn out useful to build up a dilemma.

Consider how Bondi argues for the perfect cosmological principle and, indirectly, for his own steady-state theory of the universe:

[4] Either the laws of physics, as we have them here and now, apply everywhere and at all times, because the universe has been the same at all times and is the same everywhere, broadly speaking, or cosmology is a very much more difficult subject than I would like to tackle. [40]

Rebutting rival hypotheses. Once empirical testability is accepted as the main epistemic value of science, the best way of rebutting a hypothesis is to show that it is in conflict with accepted data. But the situation is not always so easy. During a controversy, the proponent of a hypothesis will hardly admit this conflict openly, and before surrendering to his critics, he will try to attenuate the negative consequences of his own hypothesis in many ways. For example, he will adduce that the data have been misinterpreted, or that the difficulties can be overcome, or that they are not so relevant as to require the rebuttal of the hypothesis, or that the hypothesis still has other advantages. At this point, the discussion will change in tone, and the critics will pass to other difficulties. For example, they will try to show that the hypothesis does not possess the advantages the proponent believes it has, or that it is contrary to other theses already admitted by the proponent during the discussion. In this case, he will argue *ad hominem*, and sometimes he will even try to make fun of the proponent, for example, with ridiculous analogies.

Galileo is a master of this technique. Consider, for example, the following witty remark about Simplicio's thesis that the heavens are made of impenetrable hardness:

[5] What excellent stuff, the sky, for anyone who could get hold of it for building a palace! So hard, and yet so transparent! [41]

Darwin makes use of the same technique. A serious objection to his theory comes from the existence of organs in one species that seem to be detrimental to that species and useful only for another. He admits that "if it could be proved that any part of the structure of any one species had been formed for the exclusive good of another species, it would annihilate my theory, for such could not have been produced through natural selection."[42] How then can one explain the rattle of the rattlesnake? Darwin has no better answer than this:

> [6] It is admitted that the rattlesnake has a poison-fang for its own defence and for the destruction of its prey; but some authors suppose that at the same time this snake is furnished with a rattle for its own injury, namely, to warn its prey to escape. I would almost as soon believe that the cat curls the end of its tail when preparing to spring, in order to warn the doomed mouse.[43]

When he has no better weapons, the critic may attack the person of the opponent by insinuating concealed motives. Psychoanalysis (even before Freud) sometimes serves this purpose well. Consider how Galileo attacks Simplicio again on the question of the incorruptibility of the heavens:

> [7] Those who so greatly exalt incorruptibility, inalterability, etc. are reduced to talking this way, I believe, by their great desire to go on living, and by the terror they have of death. they do not reflect that if men were immortal, they themselves would never have come into the world. Such men really deserve to encounter a Medusa's head which would transmute them into statues of jasper or of diamond, and thus make them more perfect than they are.[44]

A barrister behaves no differently when during a cross-examination, he starts by questioning the deeds perpetrated by the accused, goes on to attempt to show the implausibility of what he admits, and finally tries to discredit him by throwing unfavorable light on certain of his habits or personal inclinations.

Attributing plausibility to a hypothesis. A hypothesis with a low degree of prior probability (plausibility) does not have much chance of being taken into consideration. Moreover, if this degree is zero, the degree of confirmation is also zero. To attribute finite plausibility to a hypothesis is therefore essential. Such an attribution can be made in several ways. One can show that the hypothesis is a deductive consequence of premises already accepted or that it inductively follows from empirical premises. But although one invokes deduction or induction, his argument is sometimes rhetorical, for the fact as to whether T' really follows from T and a premise P is a question of logic, but the fact that since T' follows from T and P, then, if T is accepted, T' is to be credited, is a question of persuasion, unless P is also proved to be acceptable. This argument is an argument from authority.

Consider the following situation: Bonnor defends his relativistic theory of the universe because it is "in satisfactory agreement with present observations,"[45] then he adds:

[8] Finally, let me stress that this theory is not constructed ad hoc to deal with cosmology. It is based on general relativity, which is known to be a satisfactory theory on a terrestrial scale and for the solar system. This gives one, I think, an added confidence in it.[46]

The crucial point here is the (deliberately) vague expression "based on." It may mean "consistent with" or "logically follows from." In the former case, it gives scarce credit to the conclusion; in the latter, it would render the argument compelling only if the additional premise, especially the highly debated question whether relativity theory can be extended from a terrestrial to a cosmological scale, were clearly stated and settled. As it is, the argument is a typical invocation of authority: the faith in the father, as it were, is invoked to support one of his children.

But this is not the only way of lending credibility to a hypothesis. Sometimes one can make use of analogies. Consider Darwin's hypothesis of natural selection. One of his reasons for giving it prior probability is the following:

[9] Can the principle of selection, which we have seen is so potent in the hands of man, apply to nature? I think we shall see that it can act most effectually. . . . Can it, then, be thought improbable seeing that variations useful to man have undoubtedly occurred, that other variations useful in some way to each being in the great and complex battle of life, should sometimes occur in the course of thousands of generations?[47]

A different type of analogy, an argument from hierarchy, as we could call it, allows Darwin to increase the initial degree of probability of the natural selection hypothesis. He writes:

[10] As man can produce and certainly has produced a great result by his methodical and unconscious means of selection, what may not nature effect? Man can act only on external and visible characters: nature cares nothing for appearances, except in so far as they may be useful to any being. She can act on every internal organ, on every shade of constitutional difference, on the whole machinery of life.[48]

Although analogy is recurrent, there are no typical rhetorical arguments in this context. Anyone who wants to attribute plausibility to his own hypothesis or to reinforce it may resort to many different techniques of persuasion. Sometimes he will try to show that it is of the same kind admitted or praised by his interlocutors; sometimes that similar hypotheses have proved fruitful in other fields; at other times, that the doubts and difficulties it raises can be overcome or balanced by other advantages. Thus he may invoke the same treatment conceded to others in similar situations (analogy, retortion); he may appeal to certain relevant cases taken as similar (recourse to precedents); he may invite his interlocutors to be confident and trust in future gains (pragmatical arguments); and so on.

Clearly, the contexts above do not exhaust all the functions rhetorical arguments perform in science. But they are sufficient to allow us to conclude that rhetoric in science plays a role that is not merely ornamental or embellishing. One might object that rhetoric is resorted to only when induction from evidence or deduction from principles is insufficient, at the beginning or in the middle of a dispute, as it were, not at the end, when observations and experiments have finally proved their worth. But this is not clear. When can we say that a dispute is at its end? When we possess all the relevant evidence? A remark by Bondi is apposite here: "we can never wait until we have all the facts at our disposal; that time never comes."[49] This does not mean that empirical evidence does not count or that a scientific dispute never stops or we can never get rid of it. Of course, empirical evidence is essential, but, as Aristotle said, it is also essential "to raise searching difficulties on both sides of a subject." And, of course, scientific disputes come to an end, and we finish with them, but we rationally conclude scientific disputes in the same way as we conclude any other dispute: by confronting, attacking, and persuading our interlocutors with good reasons.

III. The Basis of Scientific Dialectics

"Good" in what sense? What does "good reasons" mean in science? This is the next step of our project.

Aristotle seems to hold the view that if the *éndoxa* are shared and the syllogism that takes them as premises is valid, then the argument counts as a good reason. But the second part of this view is too narrow, for it does not apply to rhetorical arguments, which cannot be said to be valid or invalid on merely formal grounds.

Let us take, for example, Bondi's dilemma [4] about the perfect cosmological principle. In order to evaluate it, we have to consider different aspects of it. For example, we have to establish whether the two horns of the dilemma really cover the entire situation and whether there is no way of passing between them; whether the second horn really makes cosmology impossible or simply more difficult; whether such difficulties can be overcome; whether the first horn leads to other difficulties; whether these difficulties are more or less serious than the others; and so on.

Consideration of all these aspects involves discussion. But discussion pre-supposes a framework shared by the participants. So, in order to evaluate an argument, we have to examine not merely its form or structure in isolation, but the way the argument relates to the elements of this framework admitted by the audience. Suppose one says "H is true," and the reason he gives is "because H is written in the Holy Scriptures." This argument would be considered to be bad because the authority of the Scriptures does not figure among the admitted bases for scientific knowledge. Suppose one says "H is promising," and his reason is the same as Bonnor's reason in argument [8]: "because H is based on T, which is a well-accepted theory." If "based on" is taken in the sense of "follows from," this argument would be considered to be good because consistency with accepted theories is one of the values the audience recognizes. But suppose an interlocutor replies: "H is not promising because it follows from T, provided P is accepted, and P is a highly debatable assumption." At this point, a discussion arises. We cannot say in abstract terms who is right and who is wrong; there is no impartial "tribunal" to deliver a secure verdict, no "scales of proofs" weighing the merits of rival opinions, no "certain and simple rules" of method establishing truth or falsity. There is only the discussion and the abilities of the interlocutors to turn the theses conceded and the factors admitted to their own advantage.

But is not this discussion regimented by rules? It must be because we do not want to say that the argument of him who happens to win a discussion is, *eo ipso*, good or bad. We want to say the other way round, namely, that that argument that is good scores a victory in a discussion.

We then need a logic. But what kind of logic? As formal logic and inductive logic (whatever it is) are not sufficient to evaluate rhetorical arguments, we need a logic regimenting scientific discussions because, as we have seen, rhetorical arguments can be evaluated only within a discussion context. I shall call this logic *scientific dialectics*, and I shall take scientific dialectics as the canon of scientific rhetoric. To make up this canon, we have to perform two operations. We have to establish the premises of scientific discussions (the equivalent of Aristotle's *éndoxa*) and the ways in which they are conducted. I shall call the former the *substantive factors* of scientific rhetoric and the latter the *procedural factors*, and rephrasing an expression of Chaim Perelman for my own purposes, [50] I shall refer to them both as the *basis of scientific dialectics*. Table 1 contains what I think is the essential of this basis.

A detailed examination of all these factors cannot be given here. But a schematic presentation is needed.

As for the substantive factors, facts and theories are taken in the standard ways, as results of observations and experiments, and as accepted explanatory hypotheses. Facts and theories play an essential role in a scientific discussion. A

Table 1: The Basis of Scientific Dialectics

Substantive Factors	*Procedural Factors*
Facts	Rules for conducting discussion
Theories	Rules for terminating discussion
Values	
Assumptions	
Loci	
Presumptions	

theory that is patently contradicted by well-ascertained facts has little hope of enhancing its value; in the same way, a thesis that is inconsistent with accepted theories has little chance of being taken into consideration. When facts and theories are not sufficient, one can resort to assumptions, that is, ontological views both general and disciplinary, and to epistemic values. Loci are the next step; they are principles of preference that establish hierarchies of values. Lastly, we have the presumptions. Their role is fundamental for they are suppositions, valid until proved to the contrary, to the effect that a thesis that fulfills certain desiderata or possesses certain properties is taken for granted unless it is proved untenable.

As for the procedural factors, the rules for conducting scientific discussion do not differ essentially from those of ordinary discussions. In particular, they admit that an interlocutor may withdraw a thesis. This does not exclude recourse to *ad hominem* arguments; such arguments retain their efficacy when the interlocutor commits himself not to withdraw a certain thesis or when he cannot objectively withdraw it. The rules for terminating the discussion are also typical of all rhetorical contexts. For example, we say that a debate between two interlocutors A and B is lost for A if B proves his own thesis on the basis of premises put forward by A, if B reaches a conclusion that is covered by a presumption accepted by A, if A is led to contradict himself and is not able to find a solution, if A is led to deny certain of the substantive factors he himself admits, and so on.

A few remarks are important as regards these factors. The set of the substantive factors is not fixed and cannot be fixed once and for all. Let us introduce the expression *configuration of the substantive factors* to indicate a concrete set of substantive factors of scientific rhetoric accepted in a concrete situation with their corresponding criteria, interpretations, and hierarchies. Then different epochs may have different configurations. This introduces a certain relativism into science, for an argument may be good according to one configuration and bad according to another. This sort of relativism cannot be avoided unless one dreams of universal, permanent standards in the light of which arguments are good or bad *ab aeterno*. But such a relativism should not be blamed, for it reflects a situation that frequently arises. Consider, for example, Darwin's arguments for his theory of natural selection. His resistance to publication of the *Sketch* in 1842 and of the *Essay* in 1844 was overcome in the late 1850s when he realized that the scientific community had changed its mind about the theory, in particular about one of its fundamental assumptions, that is, transformism. An argument that would have been considered weak before was then regarded favorably and was slowly accepted.

However, relativism does not go too far. Although we can say that each epoch has its own configuration of factors, we cannot say that each epoch has its own factors. Configurations depend on the relative weight attributed to factors in different historical contexts, but factors depend on a *tradition*. While configurations alter the relative position of factors, tradition establishes them. Most of the factors on which current scientific research relies are those that were first established by the Greeks, and then endorsed by Galileo and his followers. So in science, we aim at agreement of knowledge claims with observations; we want accuracy, simplicity, coherence, and so on. Moreover, factors also depend on an *attitude*. Science is an attempt to understand nature, which is different from considering it from other points of view, such as religious or aesthetical. So in science, we presuppose that nature is intelligible; we assume that it has an order, and so on. This holds good for procedural factors as well. The argumentative forms of persuasion may be thought to stem from the critical tradition born in Greece, and more deeply, from the natural attitude of argumentative reasoning. Again, this does not mean that factors are fixed once and for all. Permanence is not important; what matters is continuity. Continuity of factors is compatible with different configurations of them, but it is not compatible with that sort of radical relativism according to which each epoch has its own views (sets of beliefs, conceptual schemes, truths, and so on), and there is no rational, argumentative way to discuss a change from one view to another.

Once the factors are identified, our last step is the evaluation of arguments. The notions introduced here may be used to propose some explications of the relevant evaluative expressions.

Let us start with "pertinent." This is an ambiguous notion (like "intelligent"), for it can be taken both as descriptive and evaluative. We can say that an argument is pertinent in scientific contexts if it refers to the factors of science. Thus we can suggest the following explication:

Expl. 1   A scientific argument, in a certain field for a certain function, is *pertinent* to a thesis if the reasons advanced for that thesis belong to the substantive factors of scientific dialectics admitted in that field for that function.

"Valid" (or "good") is the next notion. We have already seen that the assessment of an argument in this respect presupposes a discussion. An argument is valid if it is pertinent and its proposer successfully faces the criticism against it, that is, if he is not defeated by his interlocutors during a discussion. To define the validity of an argument, we have then to establish when an interlocutor is not defeated in a discussion. We introduce the idea of a *winning dialectical strategy* in terms of the basis of scientific dialectics as follows:

> Expl. 2   A dialectical strategy for a thesis T is winning for a part P against another part Q if, on the basis of premises conceded by Q and the procedural factors of scientific dialectics, P forces Q to assent, silence or withdrawal from the debate.[51]

This suggests the following explication for the goodness or soundness or force of an argument:

> Expl. 3   A scientific argument, in a certain field for a certain function, is *valid* (*good*) if its conclusion is supported by a winning dialectical strategy on the basis of shared premises and the substantive factors of scientific dialectics admitted in that field for that function.

Then comes "strong" ("weak"). "Strong" is different from both "valid" and "effective" because some arguments may be valid but weak, and effective without being good. We can try to capture the difference between good and strong by resorting to our notion of configuration of the substantive factors. We thus suggest:

> Expl. 4   A scientific argument in a certain field for a certain function is *strong* if its conclusion is supported by a winning dialectical strategy on the basis of shared premises and the substantive factors of scientific dialectics holding in the situation in which it is advanced.

"Effective" is a descriptive notion. An effective argument is an argument that convinces the interlocutors it is addressed to. But we can try to understand why certain arguments happen to convince while others do not, or why certain arguments are more convincing than others. The notion of configuration seems to turn out useful for this purpose, too. We may pose:

> Expl. 5   An argument is effective for an interlocutor (or an audience) *I* if the reasons supporting its conclusion belong to that configuration of the substantive factors of scientific rhetoric that *I* takes as optimal.

Explications 4 and 5 allow us to understand why an effective argument may not be strong. An argument is effective for *I* on the basis of the configuration accepted by *I*, while it is strong on the basis of the configuration accepted in the situation in which it is advanced. If the two configurations coincide, an effective argument is also strong. But they may not coincide, for example, when *I* is a person disagreeing with the prevailing configuration of his time. *Ad hominem* arguments are the most effective, for they put an interlocutor at cross-purposes with his own views, but they are not necessarily strong. The judge of efficacy is the individual; the judge of goodness is the community.

One might object that the above explications of relevance, validity, strength, efficacy do not permit precise verdicts. This is largely true. The place where an argument can be evaluated is a discussion, a debate. And like Freud's analysis, a scientific debate is *interminable*. An interlocutor bogged down in serious difficulties may find a way of getting out of them, of counterattacking, and of turning the situation to his own advantage. However, like Freud's analysis, a scientific debate is also *terminable*. Feyerabend has castigated methodology with the objection that unless methodological standards are accompanied by time limits, they are "vacuous" or merely "verbal ornaments."[52] This does not hold for the dialogical view. One can say that the limit at which the debate is settled and beyond which there is no reasonable motive for carrying it on is reached when an interlocutor, in difficulty on the basis of the factors of scientific rhetoric, is no longer able to counterattack, or repeats his own arguments, or ignores his interlocutors' reasoning.

Our project for scientific dialectics, however imperfect it may be, could now be considered complete. But we still have to render it more palatable. And the best way is to compare it with other approaches in order to see what we lose and what we can gain.

IV. The Attraction of the Rhetorical View

A philosopher who, like me, intends to sell his project for rhetoric in science is probably doomed to end his days like Willy Loman, the frustrated protagonist of Arthur Miller's play, *Death of a Salesman*.

Let us suppose he turns up at the methodologist's shop. Proud of his achievements, the self-satisfied owner will be patronizing: "Sorry, sir, we do not need any rhetoric. We have our own method. Thanks to this divine instrument, we can establish whether a theory is scientific or metaphysical, whether it is false or confirmed, whether it is progressive or regressive, whether it is rational to accept it or reject it. As you can see, we are already equipped with everything we may need."

Let us suppose our philosopher knocks at the counter-methodologist's door. He will be ill-treated: "What do you want from me?" the guy will say. "Don't you know that method is dead? Didn't they pass on the good word to you that now in science 'anything goes'? That science is merely 'routine conversation'? That science 'has no secret of its success'? Don't bother me with your new views!"

If the rhetoric salesman-philosopher addresses the scientific community directly, he will get more or less the same reactions: "Science is based on methods," the majority will tell him. "If you have problems with it, look at our work carefully and do not listen to philosophers who usually speak about things they don't know." Or: "Science has many methods," a minor group will reply. "Scientists are opportunists; they are like handymen who don't hesitate to combine any kind of procedures, just as long as they get results."

Just like Willy Loman, our philosopher will feel frustrated. The methodologist sees him as an infiltrated anarchist; the counter-methodologist takes him to be a methodologist in disguise; the scientist simply believes he is incompetent. Whom shall he turn to? Like a smoker in the U.S., he is in danger of being driven out of any place he dares to visit. However desperate his enterprise may look, he may try to sell his project for rhetoric in science by using other rhetorical devices. For example, he may appeal again to Aristotle and remind his interlocutors that virtue lies somewhere between. If he opens a breach in their hearts, he may then have a chance of illustrating the advantages of his own view compared to their projects.

The methodological project is untenable. Not only are there no impersonal and universal rules of inquiry, but even if we take up a more modest position and consider local rules of method, it is easy to show that they cannot act as neutral yardsticks. Rules of method may be taken as Kant's imperatives of prudence;[53] but they are as numerous and as vague as these, first, because the epistemic ends or values of science may vary from time to time, and second, because the same ends or values may receive different interpretations and therefore there may be different means of attaining them. When an interpretation is at stake, only a discussion within a framework of accepted factors may solve the problem.

The counter-methodological project, be it anarchist, hermeneutic, sociological, or whatever, is also untenable. We agree that "the idea of a method that contains firm, unchanging and absolutely binding principles for conducting the business of science meets considerable difficulty when confronted with the results of historical research";[54] that this idea is "unrealistic" and even "pernicious"; that when rejecting method we also have to reject the view that "following that method will enable us to penetrate beneath the appearances and see nature 'in its own terms' ";[55] that historical, cultural, personal factors enter into science through the open texture of methodological rules. But from this it does not follow either that scientific arguments are to be replaced by "means other than arguments"[56] or that the question as to whether it is rational to accept a theory is "out of place"[57] or that "scientific theories, methods and acceptable results are social conventions."[58]

Both the methodological and the counter-methodological projects are subject to the same drawback, which elsewhere I have called the "Cartesian syndrome."[59] All of them, implicitly or explicitly, make the assumption that either science is guided by method, or it is irrational. Descartes takes his rules of method as "inborn principles."[60] Galileo believes that nature is either a book of mathematics, like Euclid's *Elements*, or a book of fiction, like Homer's *Iliad*;[61] accordingly, our way of reading it is either that of necessary demonstrations, or that of "altercations."[62] And present day methodologists, like Popper or Lakatos, and counter-methodologists, like Feyerabend or Rorty, still hold the same opinion.

But this opinion cannot be maintained. Aristotle had already remarked that "the minute accuracy of mathematics is not to be demanded in all cases, but only in the case of things which have no matter,"[63] and clearly affirmed that the alternative to mathematical rigor is not poetry but a less cogent way of demonstration.[64] The advantage of the rhetorical project is then considerable: it lets us outline a form of rationality that does not clash either with the Scylla of certainty, or with the Charybdis of subjectivity, a form that is less stringent than the formal or demonstrative one but more rigorous than that of poetry, ethics, politics, and so on. According to the rhetorical project, to be rational in science is neither to abide by a fixed code of rules nor merely "to pick the jargon of the interlocutor."[65] Rather, to be rational is to engage a dispute with interlocutors and to gain their assent. More generally, to be rational is to accept those theories, to work out those programs, to take those decisions that are supported by good reasons, "good" in the sense that they won a victory in a concrete debate conducted according to a concrete configuration of the basis of scientific dialectics.

If the rhetorical project now looks a little more attractive, many questions may still be asked of the Willy Loman who is peddling it. In particular, the following: "Willy, science is a cognitive enterprise. It aims at knowing how the world really is, at true descriptions and explanations. What has carrying off a victory in a debate to do with truth or other epistemic properties?" Willy has been asked this tremendous question since Plato's times. He has a timid answer, but has to forego it because he is aware that what Arthur Miller said of a play holds good for a paper, too, that is, that it has to respect at least the laws of physiology.

The curtain falls, and Willy resigns himself, misunderstood, to his lonely fate.[66]

NOTES

1. *Topica*, 101a 37–101b 4. Eng. trans. by W. A. Pickard-Cambridge, in *The Works of Aristotle*, Sir D. Ross, ed., Vol. I (Oxford: Clarendon Press, 1928).

2. *Topica*, 101 a 34–36.

3. *Rhetorica*, 1355a 27–34. Eng. trans. by W. R. Roberts, in *The Works of Aristotle*, Vol. XI.

4. On this distinction, see E. E. Ryan, *Aristotle's Theory of Rhetorical Argumentation* (Montreal: Les Editions Bellarmin, 1984), especially pp. 40–47 and 55–77.

5. *Rhetorica*, 1355a 5.

6. *Rhetorica*, 1354a 1.

7. *Metaphysica*, 1025b 12–13. Eng. trans. by Sir D. Ross, in *The Works of Aristotle*, Vol. VIII.

8. *Analytica posteriora*, 71b 18. Eng. trans. by G. R. G. Mure in *The Works of Aristotle*, Vol. I.

9. Ivi, 90b 23.

10. Ivi, 100b 3ff.

11. Recently this has convincingly been argued by Berti. He stresses that Aristotle's physics "starting with phenomena, confronts both sensible observations and most credited opinions," and he goes so far as to write that "Aristotle's demonstrations based on supposed empirical observations, like those concerning the famous 'natural places,' are less valid than those based on simple dialectical reasonings." See E. Berti, *Le ragioni di Aristotele* (Rome-Bari: Laterza, 1989), p. 59. Mansion had noted that Aristotle's procedure in physics and cosmology does not correspond to his theory of scientific demonstration in *Posterior Analytics*; he called Aristotle's method in these sciences a "bastard method" or a "hybrid method." See A. Mansion, *Introduction à la Physique Aristotélicienne* (Louvain: Editions de l'Institut Supérieur de Philosophie, 1946), p. 211.

12. *De caelo*, 279b 4–7. Eng. trans. by J. L. Stocks in *The Works of Aristotle*, Vol. II.

13. *De caelo*, 294a 10–297a 6.

14. See, for example, *De caelo*, 293a 25–27, where Aristotle blames the Pythagoreans, for "they are not seeking for theories and causes to account for observed facts, but rather forcing their observations and trying to accommodate them to certain theories and opinions of their own."

15. See the following passage from *De caelo*, where Aristotle rejects the view of Thales and his followers that the earth floats on water: "These thinkers seem to push their inquiries some way into the problem, but not so far as they might. It is what we are inclined to do, to direct our inquiry not by the matter itself, but by the views of our opponents: and even when interrogating oneself one pushes the inquiry only to the point at which one can no longer offer any opposition. Hence a good inquirer will be one who is ready in bringing forward these objections proper to the genus, and that he will be when he has gained an understanding of all the differences" (*De caelo*, 294b 6–13). See also Saint Thomas's comment upon this passage: "If one wishes to find a true solution he must not content himself with the objections he has at hand. As he

himself [Aristotle] says, the person who wants to search for the truth must be ready to consider what impinges on him and others; not through sophistic problems, but through problems which are real, rational, proper, that is to say, convenient to the genus into which he is inquiring." Cf. S. Thomae Aquinatis, *In Aristotelis libros De Caelo et Mundo, De Generatione et Corruptione, Meteorologicorum Expositio*, R. M. Spiazzi O.P., ed. (Turin: Marietti, 1952), n. 503. Commenting upon the same passage, Elders writes that "apparently scientific study is considered a dialectical enterprise." See L. Elders, *Aristotle's Cosmology. A Commentary on the 'De Caelo'* (Assen: Van Gorcum, 1965), p. 250.

16. F. Bacon, *Novum Organon*, in *The Works of Francis Bacon*, J. Spedding, R. L. Ellis, D. D. Heath, eds. (London: Longman, 1860), Vol IV, p. 42 (Preface).

17. F. Bacon, *Novum Organon*, p. 52 (§ 29).

18. G. Galilei, *Dialogue Concerning the Two Chief World Systems*, trans. by S. Drake (Berkeley: University of California Press, 1953), p. 54.

19. *Ibid.*, p. 268.

20. See G. Galilei, *Esercitazioni filosofiche di Antonio Rocco con Postille di Galileo*, in *Opere*, Edizione Nazionale (Florence: Barbèra, 1965), Vol. VII, pp. 731–734. For the method of mechanics, see also Galileo's early essay *Le mecaniche*, in *Opere*, Vol. II, p. 159.

21. The difference between procedures C and D is probably the same as that between hypothetic method (*ex suppositione*) and retroductive method, according to McMullin's reconstruction. See E. McMullin, "The Conception of Science in Galileo's Work," in *New Perspectives on Galileo*, R. E. Butts and J. C. Pitt, eds. (Dordrecht: Reidel, 1978), pp. 209–257. There is no clear indication in Galileo about the nature of the first link in sequences C and D (the other two links are as in Aristotle's A). If we look at his jargon, we may suppose that the passage from observation to axiom or hypothesis is not, according to him, inductive in a strict sense, but alogical. In his *Postille* on Rocco's *Esercitazioni*, he uses such expressions as "I made up an axiom" (*mi formai un assioma*), and "I framed in my mind" (*mi figurai con la mente*) that seem to allude to free associations, rather than logical connections.

22. See Galileo's *Trattato della sfera*, in *Opere*, Vol. II, pp. 211–12, where the four steps of this procedure are clearly stated and numbered.

23. Hempel is right in remarking that "the problem of formulating norms for the critical appraisal of theories may be regarded as a modern outgrowth of the classical problem of induction." See C. Hempel, "Valuation and Objectivity in Science," in *Physics, Philosophy and Psychoanalysis*, R. S. Cohen and L. Laudan, eds. (Dordrecht-Boston: Reidel, 1984), pp. 73–100 (p. 92).

24. In *De caelo*, 279b 10–13, Aristotle depicts his dialectical procedure in juridical terms: "to give a satisfactory decision as to the truth it is necessary to be rather an arbitrator than a party to the dispute." Kant took his critique of reason as a "tribunal."

See *Critique of Pure Reason*, trans. by N. Kemp Smith (London: MacMillan, 1978), A xi–xii. The metaphor of experience as a tribunal or a judge has been used by many empiricists. See W. V. O. Quine, *From a Logical Point of View* (Cambridge, Mass.: Harvard University Press, 1961, 2nd ed.), p. 41; K. Popper, *The Open Society and its Enemies* (London: Routledge and Kegan Paul, 1966, 5th ed.), p. 218; *The Logic of Scientific Discovery* (London: Hutchinson, 1959), p. 109.

25. In a private communication, Philip Kitcher rightly reminds me that according to Quine our knowledge depends on information acquired from historical tradition. This is true, as it is true that Quine, quite consistently with this view, conceives of physical objects as "cultural posits." However, Quine seem still to consider scientific knowledge as a game with two players: the community of inquirers, which is a single, collective subject, on the one hand, and the "tribunal of sense experience," on the other. If the players are more than two, experience is no longer a tribunal, a monocratic judge, but a member of a jury together with other members.

26. F. Bacon, *Novum Organum*, p. 63 (§ 61; see also § 122).

27. R. Descartes, "Rules for the Direction of the Mind," in *The Essential Descartes*, M. D. Wilson, ed. (New York: New American Library, 1969), p. 44.

28. G. W. Leibniz, *Philosophische Schriften*, hrsg. C. I. Gerhardt, 7 vols. (Hildesheim: G. Olms, 1961), Vol. VII, p. 202.

29. K. Popper, *The Logic of Scientific Discovery*, cit., p. 39.

30. I. Lakatos, "Why Did Copernicus's Programme Supersede Ptolemy's?" in I. Lakatos, *Philosophical Papers*, 2 vols., J. Worrall and G. Currie, eds. (Cambridge: Cambridge University Press, 1978), p. 192.

31. I. Lakatos, "The Role of Crucial Experiments in Science," *Studies in History and Philosophy of Science* 54 (1974), p. 315n.

32. Of his science of method, Descartes writes that "such a science should contain the primary rudiments of human reason." See Descartes, "Rules," p. 46.

33. Galileo, *Dialogue*, p. 103. See also Descartes, "Rules," p. 45: "For the human mind has in it [*sc.*: method] something that we may call divine, wherein are scatterd the first germs of useful modes of thought."

34. *De caelo*, 296a 34–b 6.

35. *The Autobiography of Charles Darwin*, N. Barlow, ed. (New York: Norton & Co., 1958), p. 119.

36. C. Darwin, *The Origin of Species* (London: Everyman's Library, 1982), p. 455.

37. See W. Hopkins, "Physical Theories of the Phenomena of Life," in *Darwin and his Critics*, D. L. Hull, ed. (Chicago: The University of Chicago Press, 1973), p. 231.

38. Bondi *et al., Rival Theories of Cosmology* (Oxford: Oxford University Press, 1960), pp. 17–18.

39. Ivi, p. 11.

40. Ivi, p. 38.

41. Galileo, *Dialogue*, p. 69.

42. C. Darwin, *On the Origin of Species. A Facsimile of the First Edition*, E. Mayr, ed. (Cambridge, Mass: Harvard University Press, 1964), p. 201.

43. *Ibid.*

44. Galileo, *Dialogue*, p. 59.

45. Bondi *et al., Rival Theories*, p. 36.

46. *Ibid.*

47. C. Darwin, *On the Origin of Species*, p. 80.

48. C. Darwin, ivi, p. 83.

49. H. Bondi, *The Universe at Large* (London: Heinemann Educational Books, 1960), p. 35.

50. See Ch. Perelman and L. Olbrechts-Tyteca, *The New Rhetoric: A Treatise on Argumentation*, Eng. trans. by J. Wilkinson and P. Weaver (Notre Dame: University of Notre Dame Press, 1969), Part II.

51. This explication stems from a definition of validity originally put forward by Paul Lorenzen and then elaborated upon in E. M. Barth and E. C. W. Krabbe, *From Axiom to Dialogue. A Philosophical Study of Logics and Argumentation* (Berlin-New York: Walter de Gruyter, 1982).

52. P. Feyerabend, "Consolations for the Specialist," in P. Feyerabend, *Philosophical Papers*, 2 vols. (Cambridge: Cambridge University Press, 1981), Vol I, p. 148.

53. I construe them as imperatives of prudence rather than as imperatives of skill because in scientific matters the end of knowing what the world is like is the counterpart of what Kant takes, in practical matters, as the "one end that can be presupposed as actual in all rational beings." See I. Kant, *Groundwork of the Metaphysic of Morals*, H. J. Paton, ed. (New York: Harper Torchbooks, 1964), p. 83.

54. P. Feyerabend, *Against Method* (London: New Left Books, 1975), p. 23.

55. R. Rorty, *Consequences of Pragmatism* (Brighton: The Harvester Press, 1982), p. 192.

56. P. Feyerabend, *Against Method*, p. 153.

57. R. Rorty, *Philosophy and the Mirror of Nature* (Princeton: Princeton University Press, 1979), p. 331.

58. D. Bloor, *Knowledge and Social Imagery* (London: Routledge and Kegan Paul, 1976), p. 37.

59. See M. Pera, "From Methodology to Dialectics. A post-Cartesian Approach to Scientific Rationality," in *PSA 1986*, Vol. 2, A. Fine and M. Forbes, eds. (East Lansing, Mich.: Philosophy of Science Association, 1987), pp. 359–74; "Breaking the Link between Methodology and Rationality: A Plea for Rhetoric in Scientific

Inquiry," in *Theory and Experiment*, D. Batens and J. P. van Bendegem, eds. (Dordrecht: Reidel, 1988), pp. 259–76; "Methodological Sophisticationism: A Degenerating Project," in *Imre Lakatos and Theories of Scientific Change*, K. Gavroglu, Y. Goudaroulis, and P. Nicolacopoulos, eds. (Dordrecht: Kluwer Academic Publishers, 1989), pp. 169–87.

60. R. Descartes, *"Rules."* p. 45.

61. G. Galilei, *The Assayer*, in *Discoveries and Opinions of Galileo*, S. Drake, ed. (New York: Doubleday, 1957), p. 237.

62. Galileo, *Il Saggiatore*, in *Opere*, Vol VI, p. 245.

63. *Metaph.* III, 995a 15–16. Aristotle goes on to say that "hence its method is not that of natural science; for presumably the whole of nature has matter," which confirms that not procedure A but procedure B is the method of empirical sciences.

64. Cf. *Metaph.* III, 995a 6–12: "Thus some people do not listen to a speaker unless he speaks mathematically, others unless he gives instances, while others expect him to cite a poet as witness. And some want to have everything done accurately, while others are annoyed by accuracy, either because they cannot follow the connexion of thought or because they regard it as pettiforgery."

65. R. Rorty, *Philosophy*, p. 318.

66. This paper relies on my forthcoming book *Science and Rhetoric* to which I refer for a more analytical and comprehensive examination of several parts that I could not expand here. I greatly benefitted from comments, criticism, and suggestions by Peter Machamer and Philip Kitcher. I thank them very much for taking Willy seriously. But not even the rhetoric of the most skilled actor can transform an admission of help into a call for complicity.

Rhetoric and Theory Choice in Science

ERNAN McMULLIN

1. Neapolitan Prologue

In the night of thick darkness, enveloping the earliest antiquity, so remote from ourselves, there shines the eternal and never-failing light of a truth beyond all question: that the world of civil society has certainly been made by men, and that its principles are therefore to be found within the modifications of our own human mind. Whoever reflects on this cannot but marvel that the philosophers should have bent all their energies to the study of the world of nature, which, since God made it, He alone knows; and that they should have neglected the study of the world of nations, or civil world, which, since men had made it, men could come to know.[1]

The words are, of course, those of Vico, the first to propose a close link between rhetoric and science. As a professor of rhetoric at the University of Naples, Vico might have restricted himself to the techniques of oratory and eloquence, the customary domain of the rhetoricians of his dày. But from the beginning of his career, Vico was convinced that rhetoric, the *ars topica* of Aristotle, could properly claim a larger goal, a goal that had been almost entirely lost from sight as rhetoricians withdrew more and more into theories of ornamentation. In his early work, *Institutiones Oratoriae* (1711), rhetoric appears almost as a general theory of argumentation; it is to be concerned in large part with the invention, presentation, and evaluation of argument.[2] Its aim is, of course, persuasion, but effective persuasion also involves the search for truth. Philology is to be subordinated to philosophy, matters of style to models of reasoning.

55

At the beginning of the *Rhetoric*, Aristotle announced that: "the technical study of rhetoric is concerned with the modes of persuasion. Now persuasion is a sort of demonstration, since we are most fully persuaded when we consider a thing to be demonstrated."[3] Logic is thus a tool, but only one tool, of the rhetorician. Vico, taking his lead from Aristotle, did not oppose rhetoric and logic, as the later humanistic tradition had tended to do. He warned, however, of the dangers of too early an emphasis on formal logic in the curriculum. Critical thinking based on logic alone can be distinctly harmful because it clears away not only falsity but also probability:

> It is a positive fact that, just as knowledge originates in truth and error in falsity, so common sense arises from perceptions based on verisimilitude. Probabilities stand, so to speak, midway between truth and falsity. . . . Consequently, since young people are to be educated in common sense, we should be careful to ensure that the growth of common sense not be stifled in them by a habit of advanced speculative criticism.[4]

It is a great mistake, then, to think that:

> without any previous training in the *ars topica*, any person will be able to discuss the probabilities which surround any ordinary topic, and to evaluate them by the same standard employed in the sifting of truth. . . . Those who know all the *loci*, the different lines of argument, [on the other hand] are able . . . to grasp extemporaneously the elements of persuasion inherent in any question or case.[5]

Does this mean that the rhetorician has to have some knowledge of all the sciences? What does a grasp of the *topoi* of argumentation entail? Vico maintains that the rhetorician is required to be "well versed in all fields of knowledge," that he must know "the principles of the various sciences and arts."[6] He calls on the authority of Bacon to support his own claim that no student may presume to study eloquence (rhetoric) "unless previously trained in all sciences and arts."[7] This is a far cry from the notions of oratory that he had inherited, according to which elegance of expression and effectiveness of emotive appeal were all, and such a knowledge of principles would have been deemed *de trop*.

Vico finds in Cartesian physics a paradigm instance of inappropriate formalization, a reliance on formal methods when persuasion is what is called for. He waxes sarcastic about "the tremendous debt of gratitude to those geniuses who have freed us from the burdensome task of speculating about nature." If indeed it were true that the cosmos is as they describe it, we would "owe them fervent thanks." But if it is not—and Vico is quite sure that it is not:

Let our enthusiasts pause and repeatedly ponder whether they are not carelessly following an unsafe path, leading away from the goal of the solution of the problems of nature. . . . In the geometrical field, these deductive methods are excellent ways and means of demonstrating mathematical truths. But, whenever the subject-matter is unsuited to deductive treatment, the geometrical procedure may be a faulty and captious way of reasoning. . . . The principles of physics which are put forward as truths on the strength of the geometrical method are not really truths, but wear a semblance of probability. The method by which they were reached is that of geometry, but physical truths so elicited are not demonstrated as reliably as are geometrical axioms. We are able to demonstrate geometrical propositions because we create them.[8]

But we do not create the physical natures of things, which are modelled after the archetypes in the mind of God. We should thus curb the presumption implicit in the deductivist approach to physics. This approach could work only if the natures of things were transparent to us, as they are to God, who created them *ex nihilo*. Since they are not, we have to be content with probability. Vico does not say outright that physics is therefore the domain of the rhetorician, though that seems to be the conclusion towards which he is pointing. Nor does he give any hint of the sort of alternative to Cartesian physics the rhetorician might devise.

Except one. Cartesian physics, he says:

moves forward by a constant and gradual series of small, closely concatenated steps. Consequently, it is apt to smother the student's specifically philosophic faculty, i.e., his capacity to perceive the analogies existing between matters lying far apart and, apparently, most dissimilar. It is this capacity which constitutes the source and principle of all ingenious, acute, and brilliant forms of expression. It should be emphasized that tenuity, subtlety, delicacy of thought, is not identical with acuity of ideals. . . . Metaphor, the greatest and brightest ornament of forceful, distinguished speech, undoubtedly plays the first role in acute, figurative expression.[9]

The capacity for perceiving analogy, for creating lively metaphor, is therefore the specifically "philosophical" faculty; its nurture is the task of rhetoric, not logic, since it requires the fortifying of imagination and memory. Class concepts do not lie ready to hand; they are not a simple matter of abstraction. The "acuity of ideas" that metaphor enables the philosopher to achieve differs fundamentally, it would seem, from the clarity and distinctness on which Descartes had set such store in his mechanics.[10] How would this acuity actually function in natural science? Vico does not say.

Indeed, Vico has little interest in attempting an answer to questions such as these. "The greatest drawback of our educational methods," he dryly notes, "is that we pay an excessive amount of attention to the natural sciences."[11] By the time he wrote the *Scienza Nuova*, did he think that a science of physics was in fact possible?[12] Had he conceived of a weaker sense of science, as others had already done, one that would function in the "twilight of probability," as Locke called it? It is not clear that he had. What *is* clear is that when he speaks of a "new science" of human history in his masterwork, the *Scienza Nuova* (first edition, 1725), he intends the term, 'science,' to convey the qualities of universality and necessity it had traditionally done, though the *source* of these qualities had shifted from object to subject.[13]

The terminology, in his words, of "axioms, definitions, and postulates that this science takes as elements from which to deduce the principles on which it is based and the method by which it proceeds"[14] was the familiar one of the Aristotelian and Cartesian traditions. His scrutiny of the institutions in which all men agree provides him, he asserts, with "the universal and eternal principles such as every science must have."[15] And the methodology that makes this possible derives from "a truth beyond all question," the *verum/factum* principle.[16] Phrases, such as 'indubitable principle,' 'logical proof,' 'follows with necessity,' 'establish,' 'demonstration,' are dotted through the pages of the *Scienza Nuova*.

It is worth stressing this, lest anyone be tempted to conclude that Vico was an early fallibilist, or that he had already developed a probabilistic notion of science, or that he saw in rhetoric a theory of argumentation, pure and simple. If we find in his work all sorts of resonances with recent themes in the philosophy of science, it is important to remember that he was still an eighteenth-century professor of rhetoric.[17] He says about rhetoric (eloquence), for example, that it:

> does not address itself to the rational part of our nature, but almost entirely to our passions. The rational part in us may be taken captive by a net woven of purely intellectual reasonings, but the passional side of our nature can never be swayed and overcome unless this is done by more sensuous and materialistic means. The role of eloquence (rhetoric) is to persuade; an orator is persuasive when he calls forth in his hearers the mood which he desires.[18]

Passages such as this, recall the more traditional conception of rhetoric as relying upon appeals to emotion or to elegance of expression, rather than to argumentative force.

There is, it would seem, a tension in Vico's own thinking between two rather different conceptions of the relationship between rhetoric and science. According to one, rhetoric is other-directed; it aims to persuade by whatever means are most effective to that end. It deals with the plausibility of specific arguments only because the more plausible an argument is, the more likely, on the whole, it is to be persuasive. It aids in the construction of metaphor, just as it does with other literary devices that give pleasure to the hearers and thus make it more likely that they will be persuaded in the desired way. According to the other conception, rhetoric is already partially epistemic in character. It provides an account of probable argument that logic lacks. And it aids the philosopher to hit on the metaphoric language appropriate to a science of nature. Here the stress is not so much on the persuasion of others, as on the making of the best possible case, that is, the case that comes as near the truth as possible.

In the classical works on oratory, speakers were enjoined to open with a *captatio benevolentiae*, a story or illustration that would gain the goodwill and the active interest of the audience. Has my choice of Vico been rhetorically effective? Has it made a valid philosophical point? Here we catch once again the sort of ambivalence we have already met in Vico, the tension between two rather different notions of "good" rhetoric, or if you prefer, between two different ways of relating rhetoric to the search for truth. As the reader will know, these very differences have manifested themselves in recent discussions in philosophy of science, so to them I now turn.

II. The Community of Science

Kuhn was not the first, of course, to underline the fact that science is not a solitary affair; it is a community enterprise in which the mechanisms whereby consensus is brought about are of central importance. But it was Kuhn, more than any other, who forced philosophers of science to explore the consequences of this fact in intensive and disputed detail. Paradigms, whatever else they may entail, certainly involve group commitments. From Kuhn's point of view, Newton's *Principia* did not attain the status of a paradigm until it had become the locus of authority in mechanics for the substantial majority of those working in that field. How did that come about? How in general *does* a particular achievement take on this kind of authority? Well, that, of course, is what Kuhn's book is all about. It might be carrying matters too far to describe it as a manual of rhetoric, but there can be no doubt that rhetorical concerns manifest themselves on almost every page. [19]

There are three contexts in particular where persuasion is critical if one views science as in some basic sense a communal enterprise. The first is to persuade others of the merits of one's own paradigm, or in a more restricted sense, of one's own theory. Kuhn had much to say about this. Indeed, the central thesis of his book was that the form this persuasion takes is not at all of the logically cogent sort that earlier logicist views of science seemed to take for granted. The second context is that of teaching, of introducing newcomers into the ways of the scientific community. Kuhn's most distinctive comments, perhaps, bore on the role of exemplars, of canonical scientific achievements and of approved problem-solutions, in leading apprentice scientists to see the world the way their mentors do. The third context is that of experiment, the determination that an experiment has given a reliable result on which the community can depend. This has of late become a popular topic; Peter Galison's book, *How Experiments End*,[20] may be taken as a paradigm of a growing genre.

In each of these contexts, philosophers and historians have been at pains to try to determine how persuasion typically goes on, and in particular what the balance is between epistemic and nonepistemic factors.[21] Part of the debate, of course, has been about this distinction itself: can one really separate off the epistemic (or truth-making) factors from the other factors that typically play a role in persuasion, as classical philosophy of science has always supposed one could?[22] In the metaphor of Francis Bacon, can one really decide which are idols and which are the truth-finding norms appropriate to "good" science? In the language of Hilary Putnam, is the consensus brought about by the techniques of persuasion within the scientific community itself constitutive of the truth of theory, or are there conditions on how the persuasion itself is brought about? These questions can be reformulated as questions about the role of rhetoric in science.

Why is this way of putting those much-debated questions so unfamiliar? One rarely encounters a reference to rhetoric in contemporary writings on the philosophy of science. One reason might be the ambiguity of the notion of rhetoric itself, an ambiguity we have already encountered in our brief foray into Vico's world. The term has a broad spectrum of senses, ranging from the approving to the pejorative. It is thus a risky term to employ in philosophical analysis, since it carries so heavy an emotive freight. Its use might itself be regarded as rhetorically unwise. Nonetheless, in the spirit of Vico, it seems worth the trial.

But first the ambiguity of the term 'rhetoric' must be dealt with directly. It seems to be employed in three broadly different ways. Rhetoric may be *contrasted* with logic, as it so often is in contemporary usage, so that a "rhetorical device" becomes something that works to persuade by means of something *other* than its force as legitimate argument. In this narrower, more or less pejorative sense, the rhetorical becomes synonymous with the nonepistemic. And the task of the philosopher of science becomes the separating off of logical from "rhetorical" components in the process of consensus making. We can call this the P (i.e., pejorative) sense. To call a piece of discourse "rhetorical" in this sense is to imply that it is not functioning as it should; it is calling on passion, social interest, or the like, in ways that may be effective in persuasion, but that do not further the truth (presumed to be the primary aim of the discourse).

Rhetoric can also be used in the opposite way, limiting it to the argumentative aspect of discourse only, so that a "good" piece of rhetoric would be a good argument as well as an effective instrument of persuasion. Here, rhetoric and logic are not opposed but are in fact closely linked, so that we might call this the L sense of the term. This is a much more limited sense than the first; it can be found, for example, in courses of "rhetoric" along broadly Aristotelian lines. By contrast with the P sense, it is approving or at least becomes approving when used in a comparative way: "the better the rhetoric, the better the argument."

Herbert Simons conveys the contrast between these two senses particularly well:

> Because 'rhetoric' tends to be a "devil" term in our culture—often preceded by 'mere,' 'only,' 'empty,' or worse—scientists understandably recoil from it, insisting instead that their discourse is purely "objective." Yet in the classical, non-pejorative sense, 'rhetoric' refers to reason-giving activity on judgmental matters about which there can be no formal proof. The classical conception permits and even encourages the eulogistic sense of rhetoric as *good* reason-giving on matters of judgment. In the final analysis that is what defenders of science *mean* by objectivity.[23]

There is a third sense, which is relatively neutral (the N sense), where "rhetoric" refers to effective persuasion generally, whether this effectiveness derives from cogency of argument or from emotive devices or the like. In this case one would have to distinguish within the "rhetorical" techniques between the properly epistemic (those that bear on the truth of the matter) and the nonepistemic (those that work their effect independently of the truth of the matter). The distinction is not always easy to draw in practice, and indeed its propriety is challenged in various quarters in philosophy and sociology of science.

Rhetoric in this sense is neither opposed to science nor allied with it. One simply notes how crucial the role of effective persuasion is in scientific inquiry, and then one goes on to scrutinize any given use of "rhetoric" in a scientific context to determine whether it achieves its effect in epistemically defensible ways or not. To call such usage "rhetorical" leaves open the question of whether it helps or hinders the discourse as *science*. What it brings out is the relevance in this context of achieving consensus, of bringing over other members of the scientific community to a particular point of view, whatever be the means used.[24]

III. Rhetorical Factors in Theory Choice?

I want to focus now on just one of the three contexts listed above, the one where the role of rhetoric in science is most in evidence, the context of theory acceptance. When a scientist wishes to persuade his or her colleagues to accept a particular theory in preference to other alternatives, the standard procedure is to urge that the theory possesses certain virtues. Prominent among those, of course, is its empirical adequacy, its ability to "save the phenomena," to use a time-honored phrase. But other virtues are also in demand, such virtues as consistency, coherence, and fertility.[25] The second-order virtues, as I will call them, appear to play as important a role in persuasion as does the first-order requirement that a theory should predict accurately. This is puzzling, even troubling, to a strict empiricist or to an instrumentalist. If the only permissible warrant for a theory is the phenomena it "accounts for," that is, that are deducible from it as expected consequences, then why should one impose any extra requirements? *Ought* these additional virtues persuade? Or, since they quite evidently do, and empiricist philosophers of science have to be careful not to appear too normative in their approach to scientists, how *do* they persuade?

One popular answer is to label them pragmatic or aesthetic. The implication intended is that they do not bear on the genuine scientific credentials of a theory; they merely make it more acceptable to the scientific community on secondary nonepistemic grounds, such as ease of use or aesthetic appeal. What is intriguing is that among the logical empiricists, at least, reliance on these virtues is understood to be legitimate but somehow second-class. Some forms of persuasion, though effective, are clearly disapproved in scientific debate: overt appeal to self-interest or to political ideology, for example. But no one objects if a scientist urges the coherence of a theory as an argument in its favor. Though different scientists may estimate the weight of this factor differently, evoking it in order to bring about consensus is regarded as more or less normal.

Those who rejected the logicist model of science in the 'sixties relied on two arguments in particular. One was the normal underdetermination of theory by the evidence adduced in its favor. This argument was in no sense new. The hypothetical status of physical theory was widely accepted by the end of the seventeenth century. Once the main warrant of theory was seen to lie in its consequences rather than in its *a priori* intuitive force, it became clear that a multiplicity of alternative theories might equally well account for the same set of consequences. Efforts were made to show that in the long run all of these might be excluded except one. But most would have said (even Newton might have been brought to agree) that one cannot deduce explanatory theory straightforwardly from the phenomena. One task of the natural philosopher, then, was understood to be that of choosing among the perhaps many alternative accounts that save the phenomena.

The second reason for rejecting the logicist model was not new either, though it had enjoyed far less prominence than the first. Kuhn in particular stressed the fact that theory choice is guided in practice not by a set of rules, not by a *logic*, whether deductive or inductive, but by a diverse set of values. [26] To maximize one of these values is not to apply a rule; different members of the scientific community may understand them somewhat differently and are likely to weight them differently. Saving the phenomena is only *one* of the virtues that scientists look for in a theory, albeit in the long run (not necessarily in the short), it is the most important one. It is the *other* values, the second-order ones, therefore that are likely to be decisive if several of the competing alternatives appear more or less equally adequate from the predictive standpoint.

The first who alluded explicitly to this issue was perhaps Johannes Kepler. In his *Apology for Tycho against Ursus* (1600), he defended Tycho Brahe against the critique of Nicholas Baer (Ursus in the Latin form of his name). [27] Baer took the standard view of mathematical astronomy of that time, that its constructions ought be read only as convenient devices for prediction. Brahe was convinced, as Copernicus had been before him, that his own model of the planetary motions was a *true* account and not just an ingenious way to save the phenomena. But how was one to meet the instrumentalist objection that many different hypotheses might equally well fit the phenomena? Copernicus had hit on an argument from the "naturalness" of his own account in contrast with the *ad hoc* character of the Ptolemaic alternative. But it was Kepler who first saw, with quite surprising clarity, that such second-order considerations would be essential if the instrumentalist interpretation of astronomy were to be overcome. He argued that numerous oddities of the Ptolemaic model, the fact that each of the planets has associated with it a circular motion with a period of exactly one year, for example, find an explanation in the heliocentric system of Copernicus. This system is to be preferred, not because it predicts better, but because it *explains* better; it suggests *causes* for various peculiarities of the planetary motions that appear as *ad hoc* postulations in the system of Ptolemy. This is a second-order virtue, plainly, one that is often today called *coherence*. It carried conviction for Kepler and many of his contemporaries. [28]

It is easy to find similar cases in later science. But I want to move directly to the twentieth century in order to see what the logical positivists made of these second-order virtues. In his inductive logic, Carnap admitted only the sort of quantitative support that evidence in the form of logical consequences could offer to an hypothesis. But in response to questions, such as: do numbers, propositions, space-time points exist?, he distinguishes between "internal" questions, that is, questions that can be answered within a given linguistic framework, once that framework itself has been accepted, and "external" questions, that is, questions that bear on the acceptability of the framework itself.[29] Existence questions, such as: do numbers exist?, are external; they cannot be answered in the normal empirical, scientific way. They are in fact pseudo-questions. What one *can* meaningfully ask is the pragmatic question: what are the advantages of the language of mathematics?

Carnap went on to draw a further distinction between belief and acceptance. One can accept such entities as numbers or space-time points, in the sense of recognizing the merits of the framework in which they occur. But such an acceptance does not warrant belief in the existence of the entities:

> The acceptance cannot be judged as being either true or false because it is not an assertion. It can only be judged as being more or less *expedient, fruitful, conducive to the aim* for which the language is intended. Judgements of this kind supply the motivation for the decision of accepting or rejecting the kind of entities. . . . Thus the question of the admissibility of entities of a certain type or of abstract entities in general as designata is reduced to the question of the acceptability of the linguistic framework for those entities.[30]

Carnap was so preoccupied with abstract entities, such as numbers and propositions, that he does not appear to have adverted to the fact that *any* scientific theory can be regarded as a "linguistic framework." Hence, the features looked for in a good theory must be, even on his own account, much broader in nature than those regulated by his inductive logic. Of course, only the latter are properly "cognitive" in his view. The others are pragmatic. But since they bear on the acceptability of the framework, that is, of the theory, in the scientific community, they are in some sense prior to the matter of inductive confirmation. In the end, his view would seem to lead to a sort of instrumentalism, and indeed Carnap says: "The acceptance or rejection or abstract linguistic frameworks will finally be decided by their efficiency as instruments, the ratio of the results achieved to the amount and complexity of the efforts required."[31]

Philipp Frank addressed this issue in a much more detailed way. He argued that once one replaces the dyadic scheme of intellect (or idea) and world that characterized classical philosophy with Peirce's triadic scheme of object, sign, and sign user, one is forced to move from belief to the broader notion of acceptance in order to incorporate the pragmatic dimension.[32] And hence, in the context of scientific theory, it is not enough to rely on the usual logical and semantical criteria of consistency and agreement with the facts. These alone leave theory underdetermined. Additional pragmatic criteria are required and indeed are everywhere in evidence in the history of science:

> Today, everyone who has attentively studied the logic of science will know that science actually is an instrument that serves the purpose of

connecting present events with future events and deliberately utilizes this knowledge to shape future physical events as they are desired. This instrument consists of a system of propositions—principles— and the operational definitions of their terms. These propositions certainly cannot be derived from the facts of our experience and are not uniquely determined by these facts. If the principles or hypotheses are not determined by the physical facts, by what are they determined? We have learnt by now that, besides the agreement with observed facts, there are other reasons for the acceptance of a theory: simplicity, agreement with common sense, fitness for supporting a desirable human conduct, and so forth.[33]

Most scientists would maintain, he goes on, that such "extrascientific" reasons ought not be allowed to influence science. Their denial though in clear contradiction with the actual practice of science, is presumably motivated by their having been told that scientific theory ought to be determined by the facts only. What they must be brought to realize is that science is an instrument of prediction and control only. The additional factors that determine which theories are actually chosen do not derive from the facts but from external conditions, from moral, political, and religious considerations, for example. Theory decision is not arbitrary then, but the factors guiding the decision to accept Copernican or Darwinian theory, say, involve larger issues about the purposes of human life and the furtherance of a particular moral agenda. There is nothing wrong with this, Frank insists. These extra factors that make actual theory choice possible are legitimate in larger human terms, even though they do not bear on objective physical reality.

Might these second-order criteria be called "scientific"? Frank confesses himself puzzled to know how to respond. They are surely not scientific in the conventional sense; they allow everything from the fertility (what he calls the "dynamism") of a theory to its consistency with broadly held political or religious views to guide judgment on the theory's acceptability. But this is adjudicated by the community of science as a whole; it is thus (he suggests) subject to social science: "social science had to decide whether the life of man would become happier or unhappier by the acceptance of the Copernican system."[34] Once the pragmatic character of theory choice be recognized, he concludes, "it is difficult to draw a clear dividing line between strictly scientific and sociological criteria."[35]

It is hardly necessary to emphasize the significance of this shift. The logical analysis of natural science in syntactic and semantic categories leads Frank to a straightforwardly instrumentalist position in regard to the significance of scientific theory. But then he goes on to allow that a large-scale theory may be called on "to support desirable moral and political doctrines."[36] However, if it is no more than an instrument of prediction, how *can* it support such doctrines? Is it not reduced to the status of propaganda? Only those who are unaware of its limitations as a claim about the nature of the world could (it seems) allow themselves to be influenced by it in a broader philosophical context.

The original positivism was much too narrow; it excluded too much. But Frank's version of pragmatism is far too broad; it admits too much. It interposes no barrier to ideology; it sets up no effective truth requirements for the considerations that are permitted to govern theory choice.[37] What leads him to this awkward form of pragmatism is the manner in which he distinguishes between the "purely scientific" and the "sociological" factors in theory choice. By drawing the boundary so tightly around the first, he enlarges the second to include factors that do not belong together. The result is an impoverished instrumentalism on the one hand, and the legitimizing of ideological intervention of all sorts in science on the other. A distinction between epistemic and nonepistemic factors is entirely in order, but Frank sets the border in the wrong place. Only by reconsidering where it should be placed can the epistemic character of scientific theory and the ontological status of its constructs be secured. One must sift among the factors that Frank lumps together to discover which are legitimate criteria to impose on scientific theory and which risk being ideological intrusion.

What has all this to do with rhetoric? Though Frank does not use the term, his original point might have been made by decreeing a separation between "logic" and "rhetoric," and noting that the "sociological" or "pragmatic" considerations that he allows into theory choice will require persuasion to put into effect. They are thus rhetorical in some sense. Indeed, what has to be done is to discover in just what sense they *are* rhetorical. How do they persuade? And on what grounds? Ought they persuade? I am not going to be concerned here with detailed answers to these important questions, typical of rhetorical analysis in the older tradition. My aim in singling out Philipp Frank's work for discussion was to show how important this sort of analysis becomes as positivism shades into pragmatism.

IV. The Second-Order Virtues

The point could be made again in a more recent context. The sort of empiricism that Bas van Fraassen espouses in his influential book, *The Scientific Image*, is reminiscent of logical positivism in important ways, though the author is at some pains to separate himself from that earlier doctrine. His own "constructive" brand of empiricism is defined as holding that "science aims to give us theories which are empirically adequate; and acceptance of a theory involves as belief only that it is empirically adequate,"[38] that is, that it "saves the phenomena, correctly describes what is observable."[39] Theory acceptance may (he notes) be guided by virtues other than empirical adequacy, the only properly epistemic virtue, in the restricted sense of 'epistemic' proper to a system in which one may not inquire after the *truth* of a scientific theory. These additional virtues must be regarded as *pragmatic*, and "pragmatic virtues do not give us any reason over and above the evidence of the empirical data for thinking that a theory is true."[40] He asks: "What can an empiricist make of these other virtues," and he lists simplicity, coherence, and explanatory power, "which go so clearly beyond the ones he considers pre-eminent?"[41]

And he responds:

> There are specifically human concerns, a function of our interests and pleasures, which make some theories more valuable or appealing to us than others. . . . These factors are brought to the situation by the scientist from his own personal, social, and cultural situation. . . . They do not concern the relation between the theory and the world, but rather the use and usefulness of the theory; they provide reasons to prefer the theory independently of questions of truth.[42]

But then an obvious question occurs to him: "why is this [that is, reliance on the "theoretical" virtues, as he call them] "a *rational* procedure to follow in the appraisal of theories?" He struggles—unsuccessfully, to my mind—to answer this question and ends with an admission: "But it might be arguable that, for purely pragmatic (that is, person- and context-related) reasons, the pursuit of explanatory power [one of the theoretical virtues he lists] is the best means to serve the central aims of science."[43]

Why should it be the best means, however? Or to turn this around, if this *is* the best means to serve the aims of science, can the reasons in support of its use be "purely pragmatic (that is, person- and context-related)"? It would seem that if superior explanatory power is the mark of a theory that is in the long run more likely also to be empirically adequate (the "central aim of science" for him), it cannot be regarded as a purely pragmatic criterion. Van Fraassen is at pains to show that explaining in general is strongly context-dependent. What will serve to "explain" in one context may not suffice in a different one. But there is a "standard context" in scientific work where second-order virtues, such as coherence and fertility, are generally recognized, and where claims regarding causal explanation are not significantly person- or context-related.

More fundamentally, however, if reliance on these virtues aids in the discovery and acceptance of theories that *also* prove themselves in terms of the more limited goals sanctioned by constructive empiricism, it seems clear that the virtues must have some sort of objective basis. If they merely reflected, in van Fraassen's phrase, the "personal, social, and cultural situation" of the scientist, there is no reason why taking them seriously in deciding between, or modifying, theories, would lead to an overall improvement in predictive power. The physical world is not so easily responsive to human interest, and in any event could not simultaneously answer to the divergent "personal, social and cultural situations" of many scientists at once.

Van Fraassen, like Frank earlier, lumps together considerations other than empirical adequacy that tend to affect theory choice in science as "a function of our interests and pleasures." This allows him to deny epistemic relevance to all of them alike; none of them can have any bearing on "questions of truth." The admission that explanatory power might arguably be held to be a means towards more adequate theories in "empirical" terms is already, however, a crack in the wall. 'Pragmatic' is far too capacious a label to be of service in analysis here. Some of the factors that routinely influence theory choice assuredly do reflect personal and social interest. Sociologists of science have had much to say about these factors in late years; some of them, too, would want to extend the category of the social just as van Fraassen does that of the pragmatic, and in part toward the same end of challenging the traditional objectivist construal of how theory debates come to be resolved in science.[44]

Among the so-called "pragmatic" factors, however, some, as we have seen, stand out. Notable among them are coherence, unifying power, and fertility. There is a centuries-old tradition of allowing these values a special status in theory debate. And a case can be made, it might seem, independently even of the historical record, as to why each of these criteria would be a likely indicator of the better theory, 'better' being taken here in realist terms. The relative priority of these two types of metalevel warrant is, of course, a matter of much debate both in epistemology and in philosophy of science. If realism is a contingent thesis in the sense that a theoretical natural science supporting a realist interpretation *might* have proved impossible to construct, then the intuitive case for the epistemic status of these second-order theoretical virtues takes second place to the historical demonstration that such a science has in fact been made to work.

The reader will notice that I have not listed simplicity among the theoretical virtues, despite the prominence given it by those, like van Fraassen, who deny the epistemic relevance of these virtues generally. It is easy to see why these writers *do* emphasize it. Simplicity is much easier to cast as a pragmatic criterion, "person or context related" in van Fraassen's phrase. What appears simple to one may not appear so to another. Einstein saw the tensor expressions of his general relativity theory as excelling in their simplicity; few others would see them that way. Van Fraassen takes simplicity to be the premier theoretical virtue, and then goes on to remark: "It is surely absurd to think that the world is likely to be more simple than complicated (unless one has certain metaphysical or theological views not usually accepted as legitimate factors in scientific inference)."[45] Including simplicity as an epistemic virtue obviously makes it easier similarly to dismiss other second-order factors in equally summary fashion.

Two comments are in order. Sometimes when people speak of simplicity, they mean something like coherence. Kuhn, for example, uses the term in that way when discussing Copernicus's arguments for the superiority of his own system over Ptolemy's, and then goes on to suggest (rather as van Fraassen does) that since the appeal of such a criterion must be regarded as "aesthetic" in nature, the choice between the two systems "could only be a matter of taste, and matters of taste are the most difficult of all to debate."[46] But Copernicus himself does not mention simplicity; he argues that his system can assign the "causes" of various features of the planetary motions that Ptolemy could not explain. So he is appealing to explanatory power, not to simplicity in the aesthetic sense.[47]

When 'simplicity' is used in this latter sense, it is, indeed, of dubious epistemic weight.[48] Granted, many successful scientists have extolled its merits as a criterion of truth, Einstein and Dirac among the more famous of these. Its shortcomings are not so much that it needs a larger metaphysical context to make it seem at all plausible, but that it does not seem to have been a particularly reliable guide many times in the past.[49] Let us, then, set simplicity aside in our further discussion of the second-order values. It has too often served as distraction in the debate about the epistemic merits of theory criteria other than predictive accuracy.

Paul Churchland is one who sees these merits in very clear terms indeed. In a comment on van Fraassen's philosophy of science, he asks whether the "superempirical" virtues (as he terms them) relied on in theory acceptance are "*epistemic* virtues genuinely relevant to the estimate of a theory's truth, as tradition says, or merely *pragmatic* virtues, as van Fraassen urges"?[50] Ought empirical adequacy be regarded as "the only genuine measure of a theory's truth," casting the superempirical virtues as "purely pragmatic virtues, to be valued only for the human needs they satisfy"?

His answer is to turn van Fraassen's distinction on its head:

> Values such as ontological simplicity, coherence, and explanatory power are some of the brain's most basic criteria for recognizing information, for distinguishing information from noise. And I think they are even more fundamental values than is 'empirical adequacy,' since collectively they can overthrow an entire conceptual framework for representing the empirical facts. Indeed, they even dictate how such a framework is constructed by the questing infant in the first place.[51]

Other defenders of realism have recognized, as Churchland does, that the superempirical (or as we have called them, the second-order) virtues play a key role in the argument against instrumentalism or the sort of quasi-instrumentalist view that van Fraassen presents. Few might go so far as Churchland in locating these virtues as "the brain's most basic criteria for recognizing information." But they would agree that these criteria are not merely "pragmatic" in van Fraassen's sense, that they are not simply expressions of personal preference, that they bear on truth and not merely on utility. Further, the fact that a theory exhibits these virtues provides a *prima facie* warrant for attributing a realistic import to it. Empirical adequacy alone ("saving the phenomena") does not do this. This is, of course, why it is so important to van Fraassen in shaping the case for his antirealist position to try to deprive the second-order virtues of epistemic status.

Michael Friedman is one of those who has taken up this argument. In his *Foundations of Space-Time Theories*, he argues that unifying power is the key to the realistic interpretation of theory: "A good or fruitful theoretical structure does not serve only to provide a model for the particular phenomenon it was designed to explain; rather, in conjunction with other pieces of theoretical structure, it plays a role in the explanation of many other phenomena as well." [52] He agrees with van Fraassen in accepting the model-theoretic approach to theory but argues that "embedding" observable structure in theoretical structure affords a higher degree of confirmation in the context of theoretical unification than merely taking theoretical structure as "representative," i.e., as a predictive device. Less technically:

> I claim that the point of this kind of theoretical unification [he has offered the unification of the gas laws by the molecular model of gas as an example] is not merely aesthetic; it also results in our picture of the world being much better confirmed than it otherwise would be. A theoretical structure that plays an explanatory role in many diverse areas picks up confirmation from all these areas. . . . The theoretical description, in virtue of its far greater unifying power, is actually capable of acquiring more confirmation than is the phenomenological description. [53]

I am not going to dwell on the detail of Friedman's argument, except to note one potential weakness. According to him, theories evolve simply by *conjunction* of one domain after another. Steven Savitt has pointed out that in practice *correction* occurs in such cases as often as conjunction. [54] That is, as the reach of a theory is extended, it has to be corrected to account for the empirical data in the new domain. Thus, he concludes, Friedman's argument against van Fraassen does not go through, since the original theory is often not confirmed; rather it is refuted.

There is a way in which this objection can be met, but it involves modifying Friedman's logical notion of conjunction. (I suspect that it would also entail a rejection of the semantic construal of theory that Friedman and van Fraassen share). What happens in practice is neither conjunction nor straightforward correction but an imaginative modification of the theory, of a plausible sort. And the plausibility of this modification is enhanced by the way in which it allows new predictions to be made, which are subsequently verified, or older anomalies to be eliminated.

This involves, of course, another of the second-order virtues, fertility. What I am suggesting is that unifying power *alone* may not serve to make the case Friedman wants to make against van Fraassen's antirealism. The virtue of fertility furnishes a crucial link in the realist chain of argument.[55] And so also does coherence, the elimination of *ad hoc* features, the virtue that played so important a role in the Copernican controversy.[56] In short, these virtues must be taken to work together.[57]

And together they constitute what van Fraassen refers to as explanatory power. This is not a separate virtue, as thought it could somehow be adjudicated independently of considerations of coherence, fertility, and unification. It may be contrasted with empirical adequacy, with merely "saving the phenomena," true. But even this might be misleading if the two were somehow to be taken as disjoint. To explain is not *simply* to predict. The second-order virtues are more specifically diagnostic in practice of what we loosely call "explanatory power." But to the degree that a theory does not predict correctly, neither does it explain well. Taking explanatory power to be a "pragmatic" virtue, as van Fraassen does, might easily obscure this point.

Returning finally to the question from which this discussion began, we can see that the issue between antirealist and realist can be brought to focus by a closer inquiry into the role of rhetoric in the choice and acceptance of theory. Is reliance on the second-order virtues an example of rhetoric in the P sense, as those who describe these virtues as "pragmatic" implicitly contend? Or may we see them as epistemically legitimate, though not reducible to the simple deductive relations on which the logic of confirmation traditionally depended? In this latter case, the analysis of their force would be a matter of rhetoric in the L sense, since they support more complex forms of argument than were envisioned in classical logic. We might add finally that the separating off of these second-order virtues from social, political, and other factors that influence theory decision, that is, splitting in two the heterogeneous class of "pragmatic" factors that antirealists like to group together, is a matter for rhetorical analysis in the N sense.

Our conclusion is that the debate between realist and antirealist, which is at the center of much of the recent discussion in philosophy of science, involves rhetoric in a complex variety of ways. Whether philosophers will make use of that protean term to further the debate may be doubted. But it helps to see links between this modern controversy and a larger discussion going back through Vico to Aristotle.[58]

NOTES

1. Giambattista Vico, *The New Science*, trans. Thomas G. Bergin and Max. H. Fisch (Ithaca: Cornell University Press, 1968), par. 331; p. 96.

2. See Alessandro Giuliani, "Vico's Rhetorical Philosophy and the New Rhetoric," in *Giambattista Vico's Science of Humanity*, Giorgio Tagliacozzo and Donald Verene, eds. (Baltimore: Johns Hopkins University Press, 1976), pp. 31–46. See also Giuliano Crifo's Introduction to his new critical edition of and commentary on the *Institutiones Oratoriae* (Naples: Istituto Suor Orsola Benincasa, 1989).

3. *Rhetoric*, trans. Rhys Roberts, *The Complete Works of Aristotle*, Jonathan Barnes, ed. (Princeton: Princeton University Press, 1984), 1355a 4–6.

4. *De nostri temporis ratione studiorum*, trans. Elio Gianturco as *On the Study Methods of Our Time* (Indianapolis: Bobbs Merrill, 1965), p. 13.

5. *Ibid.*, p. 15.

6. *Ibid.*, p. 78.

7. *Ibid.*

8. *Ibid.*, p. 22–23.

9. *Ibid.*, p. 24.

10. Elsewhere in natural philosophy, Descartes concedes, albeit reluctantly, a role for metaphor and allows that the assertions the philosopher makes may not reach the certainty of science proper. But he hopes that their hypothetical status may in the long run be removed. See the *Discourse on Method*, sections 5 and 6, and the discussion in my "Conceptions of Science in the Scientific Revolution," in *Reappraisals of the Scientific Revolution*, David Lindberg and Robert Westman, eds. (Cambridge: Cambridge University Press, 1990), pp. 27–92, especially pp. 339–44.

11. *Ibid.*, p. 33.

12. In his earlier work, the *De antiquissimo Italorum sapientia* (1710), he had asserted that the method of experiment could lead physics to truth; indeed, he said, physics is *more* certain than moral philosophy because "physics considers the internal motions of bodies, which are from nature which is certain," unlike the motions of the human soul (I, 1, 2). See Eckhard Kessler, "Vico's Attempt Towards a Humanistic Foundation of Science," in *Vico: Past and Present*, G. Tagliacozzo, ed. (Atlantic Highlands, N.J.: Humanities Press, 1981), pp. 73–88, p. 87.

13. See Ernan McMullin, "Vico's Theory of Science," *Social Research*, 43 (1976), pp. 450–480, p. 458.

14. Bergin and Fisch, *New Science*, par. 41, p. 26.

15. *Ibid.*, par. 332, p. 97.

16. *Ibid.*, par. 331, p. 96.

17. See Isaiah Berlin, "Vico's Concept of Knowledge," in *Giambattista Vico*, Giorgio

Tagliacozzo and Hayden V. White, eds. (Baltimore: Johns Hopkins University Press, 1969), pp. 371–389.

18. Gianturco, *Study Methods*, p. 38.

19. A point made by, for example, Herbert Simons, "Are Scientists Rhetors in Disguise? An Analysis of Discursive Processes within Scientific Communities," in *Rhetoric in Transition*, Eugene E. White, ed. (University Park: Pennsylvania State University Press, 1980), pp. 115–30. It is worth noting that perhaps the most influential and certainly the most voluminous work on rhetoric of recent times, Chaim Perelman's *La Nouvelle Rhétorique* (Paris: Presses Universitaires de France, 1958), published a few years before Kuhn's book, makes almost no mention of the distinctive context of the scientist. Though Perelman mentions scientists among the "special audiences" who may require a more specific rhetoric because of the technicality of language they employ (section 26), he does not appear to envision the possibility that the process of science might itself depend on the achievement of consensus by rhetorical means. Many contemporary writers on rhetoric insist that rhetoric is directed only to the needs of everyday discourse and not to the situation where people are "speaking as experts to experts." See Karl Wallace, "The Fundamentals of Rhetoric," in *The Prospect of Rhetoric*, L. F. Bitzer and E. Black, eds. (Englewood Cliffs, N.J.: Prentice-Hall, 1971), pp. 3–20, p. 3.

20. Peter Galison, *How Experiments End* (Chicago: University of Chicago Press, 1987).

21. See Ernan McMullin, "The Rational and the Social," in *Scientific Rationality: The Sociological Turn*, J. R. Brown, ed. (Dordrecht: Reidel, 1984), pp. 127–163.

22. Kuhn might perhaps claim an exemption from this debate, since he long ago questioned whether the classical notion of truth applies to scientific theories in the first place (*The Structure of Scientific Revolutions* (Chicago: University of Chicago Press, 1970), p. 206). There are in the strict sense no *epistemic* factors for him. But there is, of course, in his work a weaker sense of truth as effective puzzle solving, and this sense would still allow one to draw a (weakened) distinction.

23. Simons, "Are Scientists Rhetors in Disguise?," pp. 127–128.

24. The more fashionable term, 'ideology,' has something of the same force and the same ambiguity in contemporary discussion. It too has a P sense, an L sense, and an N sense. It too can be used to raise questions about the epistemic propriety of the various factors that influence a given slice of scientific practice.

25. For a broader discussion of how views have changed in regard to the role of diverse values in theory appraisal, see Ernan McMullin, "Values in Science," *PSA 1982*, P. Asquith and T. Nickles, eds. (E. Lansing MI: Philosophy of Science Association, 1983), pp. 3–25.

26. This theme appears first in *The Structure of Scientific Revolutions*, but was developed much more fully in a later essay: "Objectivity, Value-Judgement, and Theory-Choice," in *The Essential Tension* (Chicago: University of Chicago Press, 1977), pp. 320–339.

27. See Nicholas Jardine's translation of and commentary on the *Apologia*, in Nicholas Jardine, *The Birth of History and Philosophy of Science* (Cambridge: Cambridge University Press, 1984).

28. For a fuller discussion, see Ernan McMullin, "Rationality and Paradigm-Change," to appear in a volume edited by Paul Horwich (Cambridge MA: MIT Press).

29. In a late essay, "Empiricism, Semantics, and Ontology," added as a supplement to the second edition of *Meaning and Necessity* (Chicago: University of Chicago Press, 1956), pp. 205–221.

30. *Ibid.*, pp. 214, 217.

31. *Ibid.*, p. 221.

32. Philipp Frank, *Philosophy of Science: The Link between Philosophy and Science* (Englewood Cliffs, N.J.: Prentice-Hall, 1957); see "The 'Scientific' Criteria for the Acceptance of Theories," pp. 348–354.

33. "The Variety of Reasons for the Acceptance of Scientific Theories," in *The Validation of Scientific Theories*, P. Frank, ed. (Boston: Beacon, 1956), pp. 13–14. Other papers in this volume (a collection of papers delivered at the annual AAAS meeting in 1953) are relevant to this topic, particularly those by C. West Churchman, Richard Rudner, and Barrington Moore.

34. *Ibid.*, p. 8.

35. Frank, *Philosophy of Science*, p. 354.

36. *Ibid.*, p. 355.

37. If the larger political, moral, or religious doctrines of which Frank speaks were to have some truth value antecedent to science, then consistency with these doctrines might legitimately be used as a screen, perhaps, to separate among instrumentally equivalent physical hypotheses. (This is a controversial and difficult issue. See Ernan McMullin, "Models of Scientific Inference," *CTNS Bulletin* [Berkeley: Center For Theology and the Natural Sciences], *8 (2)* (1988), pp. 1–14, pp. 12–14). But Frank as a positivist began by excluding "metaphysics," in effect by denying epistemic status to such doctrines. In that event, their status is simply that of the "desirable," in a purely emotive sense. Introducing "social science," as he does, in no way reduces the arbitrary character of the conflict between rival desires or rival group interests. Social science can do no more than attest to such conflict; it cannot resolve it.

38. Bas van Fraassen, *The Scientific Image* (Chicago: University of Chicago Press, 1980), p. 12.

39. *Ibid.*, p. 4.

40. *Ibid.*

41. *Ibid.*, p. 87.

42. *Ibid.*, pp. 87–88.

43. *Ibid.*, p. 89.

44. See Ernan McMullin, "Scientific Controversy and Its Termination," in *Scientific*

Controversies, H. T. Engelhardt and A. Caplan, eds. (Cambridge: Cambridge University Press, 1987), pp. 49–91.

45. Van Fraassen, *The Scientific Image*, p. 90.

46. Thomas Kuhn, *The Copernican Revolution* (New York: Random House, 1957), p. 172.

47. For a fuller discussion, see my "Rationality and Paradigm-Change."

48. One noted defender of realism, J. J. C. Smart, takes "simplicity and aesthetic satisfyingness" to be the hallmarks of good explanation, and hence finds himself forced to postulate that "the universe is simple" (*Our Place in the Universe* (Oxford: Blackwell, 1989), p. 61). Smart realizes how vulnerable this claim is: "I do not know how to to justify this," he goes on, and simply proposes the ontological simplicity of the world as a postulate. His conclusion: "The necessity for an ontological appeal to simplicity is what worries me most in defending realism" (p. 63). If our analysis above is correct, Smart has looked in the wrong place among the second-order virtues.

49. The issue is closely tied to that of idealization. See my "Galilean Idealization," *Studies in the History and Philosophy of Science*, 16 (1985), pp. 247–273.

50. Paul M. Churchland, "The Ontological Status of Observables," in *Images of Science*, Paul Churchland and Cliff Hooker, eds. (Chicago: University of Chicago Press, 1985), p. 41.

51. *Ibid.*, p. 243.

52. Michael Friedman, *Foundations of Space-Time Theories* (Princeton: Princeton University Press, 1983), p. 242.

53. *Ibid.*, p. 243.

54. Steven Savitt, "Selective Scientific Realism, Constructive Empiricism, and the Unification of Theories." (Paper presented at APA Pacific Division meeting, 1989).

55. See Ernan McMullin, "The Fertility of Theory and the Unit for Appraisal in Science," *Boston Studies in the Philosophy of Science* (Lakatos Memorial Volume), R. S. Cohen, et al., eds. (Dordrecht: Reidel, 1976), pp. 395–432.

56. See Michael Gardner, "Realism and Instrumentalism in Pre-Newtonian Astronomy," in *Testing Scientific Theories*, John Earman, ed. (Minneapolis: University of Minnesota Press, 1983), pp. 201–265; also Ernan McMullin, "Rationality and Paradigm-Change in Science."

57. And they do, to some extent, in Whewell's notion of *consilience*. William Harper has recently shown how examples from the history of science that illustrate consilience can be used to challenge van Fraassen's claim that the *strength* of a theoretical claim varies inversely as its *security*, or in terms of our discussion here, that the second-order virtues bear on the truth of theory. See William Harper, "Consilience and Natural Kind Reasoning," in *An Intimate Relation*, J. R. Brown and J. Mittelstrass, eds. (Dordrecht: Kluwer, 1989), pp. 115–152.

58. Besides arguing for the necessity of a broader "rhetoric" of scientific argument to

replace the narrow logic of demonstration of the philosophic tradition, Vico also proposed, as we saw, that the language of science is inescapably metaphorical, thus furnishing a second entry point for rhetoric. To develop this theme would lead us into a much longer discussion that will have to await another occasion. (See, however, a suggestion as to what direction such a discussion might take, in Ernan McMullin, "The motive for metaphor," *Proceedings American Catholic Philosophical Association*, 55 (1982), pp. 27–39).

The discussion might begin from the virtue of *fertility* mentioned above and might go on to question van Fraassen's assumption that realist and antirealist can agree in supposing that the language of science is to be literally construed (Bas van Fraassen, *The Scientific Image*, p. 10). He does allude to a third option, that "science aims to be true, properly (but not literally) construed," but dismisses this view of "theory as metaphor or simile" (*ibid.*, p. 11). One important point to note in developing the theme of theory as metaphor is that, like the modern poem, theory functions in general not as a simile but as an imaginative and only partially explored construct. Its exploration in the context of new observational data gives theory its temporal dimension, as well as furnishing the strongest argument for a realist interpretation of the most successful among the theory/metaphors.

Mnemonical Loci
and Natural Loci

PAOLO ROSSI

> Truth will sooner come out from the error
> than from confusion (F. Bacon, *Novum
> Organum*, II, 20).

I. To Give Names Is the Same as to Know

Many of the botanists of the late seventeenth and early eighteenth centuries shared the problems and the opinions of the compilers of philosophical or universal languages like John Wilkins or George Dalgarno. Their point of view was that to name is the same as to know. To know the plants, Tournefort wrote in 1694, "is the same as precisely knowing the names given to them." The notes or the characteristics of a plant must be so strictly connected with the name that it will be impossible to separate the name from the thing. The order of nature, according to Tournefort, "demands that the study of plants begin by studing their names . . . We have, so to say, the key of science when we remember the names of about six hundred genera to which it is possible to refer most part of the well-known plants."[1]

Only for practical and historical reasons did Tournefort refuse to accept the idea of a radical revision of scientific language then in use and the adoption of a philosophical language:

> If plants did not have names, it would be easier to acquire knowledge about them if they were given simple names the ending of which would indicate the relations among the plants of the same genus and of the same class. However, this would mean upsetting the entire botanical language. It was impossible to pursue this kind of exactitude at the

77

beginning of this science because names were given to plants at the
time their uses were discovered. The Ancients invented the names, but
unfortunately they were unaware of the laws of philosophical or non-
causal attribution of the names.[2]

Tournefort feared an overturn of scientific language. He believed that was
nothing to be gained by such a revolution. But, according to him, philosophical
language is a very good thing in principle. The arrangement of plants into a very
precise order does not depend on the will of men. This precise order is a natural
order. By providing plants with *insigniores notae* (essential characteristics), God
gave man the possibility of a perfect language.

Linnaeus was born in 1707, one year before the death of Tournefort. Also for
Linnaeus, "Fundamentum botaniceus duplex est: dispositio et denominatio" (the
foundation of botanic is twofold: disposition and nomination). Method, which is
the soul of science, denominates every natural object, so that object *dicat proprium
nomen suum* (can pronounce its proper name).[3] What Linnaeus proposed to do
with *plants* is precisely what John Wilkins had proposed to do with *words* in his
lengthy volume *An Essay towards a Real Character and a Philosophical Language* published in London in 1668. In Wilkins's project the object of a perfect or
philosophical language was not the vegetable kingdom, but the entire universe:

> The principal design aimed in these tables is to give a sufficient
> enumeration of all things and notions, as are to have names assigned to
> them, and withall, so to contrive these as to their order, that the place
> of everything may contribute to a description of the nature of it.
> Denoting both the General and the Particular head under which it is
> placed, and the Common Difference whereby it is distinguished from
> other things of the same kinds.[4]

"By learning the characters and the names of things (Wilkins also wrote), we
should be instructed likewise in their natures": names and nomenclatures are
conventional and, in this sense, universal language is artificial and devised by man.
But a universal language is also a philosophical language that reflects the perfect
and regular order of nature. Philosophical language requires a symmetrical regularity of the tables because the order of the tables is the same as the order of Nature.
In this sense classifications are not *invented* by man, but *discovered* by man.

II. A Boundless Quantity of Plants: the Aids to Memory

All natural objects—according to Wilkins—must be defined and collocated in a
universal order, which is reflected by the tables of encyclopaedia:

If these marks or notes could be so contrived as to have such a
dependence upon, and relation to, one another, as might be suitable to
the nature of the things which they signified and so likewise if the
names of things could be so ordered as to contain such a kind of
affinity or opposition in their letter and sounds . . . we should, by
learning the characters and the names of things, be instructed likewise
in their natures.[5]

Wilkins' emphasis on the mnemonic value of philosophical language was
very strong, and it has an important historical meaning. All the authors of the art
of memory in the sixteenth and seventeenth centuries intended to order *loci* or
places into their tables. The *loci* were meant to contain, as Giulio Camillo wrote in
his *L' idea del Teatro* (1550), "all the human concepts and all things of the world."[6]

Cyprianus Kinner (the friend and collaborator of Johannes Amos Comenius)
presented his project of an artificial language as a great aid to memory: "Which
botanist would be able to remember, among the variety of conflicting authors, the
nature and the names of all the plants?" Artificial language makes this difficult
enterprise possible, because it is like a gold chain the links of which are connected."[7]

Sixty years later, Bernard de Fontenelle seems to be replying to Kinner when
he says in the *Eloge de Tournefort*:

He introduced an order into the immense number of plants irregularly
scattered upon the earth and in the depths of the sea. He distributed
them into different species, so making it possible to remember them,
and preventing the memory of the botanists from collapsing under the
weight of a endless series of names.[8]

About the art of memory and combinatory logic from Lull to Leibniz, I
wrote a book 30 years ago, six years before the very important and rightly famous
book written by Frances A. Yates.[9] That art had for many centuries been a series of
complicated techniques to prevent mnemonic collapse. Theorized by philosophers
like Giordano Bruno, commonly used by orators and preachers, reflected
in mediaeval and Renaissance imagery in art and architecture, this art survived in
modern times under a new unsuspected form: as a *method* or an essential part of it.
This point of view is, for instance, expressed as late as 1763 by Michel Adanson
in his preface to the *Familles des plantes*:

The boundless quantity of plants began to weigh on botanists. What
memory could contain so many names? Botanists, to make their
science lighter, invented methods.[10]

The construction of universal tables of natural objects was seen as a gigantic and entirely new enterprise. The boundless quantity of natural objects and the weakness of human memory were considered in 1766 by a Russian follower of Linnaeus as the main sources of the failure of the ancients to understand nature:

> Natural science, which is truly heavenly, made little progress prior to the last hundred years. The cause of that lacuna lies in the number of natural objects . . . They are so numerous and manifold that rude antiquity esteemed that they exceed the power of human memory . . . Everybody knows that memory by itself is not sufficient for so great a number of objects. The authors of ancient times were incapable of a precise terminology not having at their disposal some order for the objects, or some system.[11]

From the middle of the sixteenth century to the end of the eighteenth century, the quantity of botanical data increased in a surprising way. In the *Herbarum Icones* by Otto Braunfels (1530), we find 258 species of plants. One hundred years later, in the *Pinax theatri botanici* (1623), Gaspar Bauhin provides a list of 6,000 species. John Ray, in 1686, catalogues 18,000 species. Daubenton, in 1715, 20,000. It was a perplexing situation that concerned all the natural kingdom. With reference to the years from 1647 to 1775, Johann Friederich Gmelin (in his German translation of Linnaeus' work) lists 27 systems for the classification of minerals that had been devised by scholars from various European countries.[12]

But the "methods" were to be very useful. The plant that Tournefort, in the middle of the seventeenth century, calls "Gramen Xerampelinum, Miliacea, praetenui, ramosaque sparsa canicula, sive Xerampelinum congener, arvense, aestivum. Gramen minutissimo semine" becomes, in the dual nomenclature of Linnaeus, "Poa bulbosa."[13] As Johannes Reftelius wrote in 1672, descriptions of plants that had previously filled numerous pages "cannot be extended beyond the noun and the adjective . . . and each word now has his proper weight."[14]

III. Three Interconnected Philosophical Traditions

In George Dalgarno's and John Wilkins' works, we find a singular mixture of themes: languages, hieroglyphs, and alphabets; chifers and secret ways of sending a message; logic and grammar; artificial, universal, or philosophical language. We also find many detailed classifications of the elements and of meteors, of liberal and mechanical arts, of minerals and metals, of plants and animals. Lastly the two works also provide dictionaries of "essential words" used in the different spoken languages.

The historian, like the anthropologist, must study unusual and obsolete ideas with the same respect that he obviously has for his own ideas. The above listed mixture of themes, which is so distant from our modern approaches, is related to a series of complicated philosophical and cultural traditions: 1) the renaissance of Ramon Lull's project for the *Ars magna* in sixteenth century thought;[15] 2) the interest, which was very strong, among the followers of the so-called "Hermetic tradition," for the ancient *art of memory*;[16] 3) the success of the concept (formulated by Francis Bacon in the *Advancement of Learning* of 1605) of the so-called *real characters*. Real characters are a conventional way of nonalphabetic writing that conveys meanings without the help or use of words. Like Chinese characters and Egyptian hieroglyphs (as they were then interpreted), real characters represent neither letters nor sounds, but words, things, notions.[17]

In each of the aforementioned interconnected traditions, the project of a new kind of universal language is strictly related to a corresponding project of a universal encyclopaedia, that is to a project of a complete and well-ordered classification of the things in the world. An encyclopaedia, as a description of universal reality, is in different ways essential: 1) to the Lullian concept of a fundamental unity of human knowledge, which corresponds to the unity of the cosmos; 2) to the complicated and baroque techniques to aid memory based on the use of images, figures, synoptic tables, and "theatres of the world"; 3) to the Lullian, Cartesian, and Baconian image of knowledge as a tree around which the various disciplines are grouped like different branches springing from the same trunk.

Incidentally, the construction of a philosophical language depends, according to Descartes, on the construction of the true philosophy (*la vrai philosophie*) since otherwise it would be impossible to enumerate all the thoughts of man and put them in order. Therefore, "Do not hope to see such a language in use because it presupposes great mutations in the order of things." Commenting on Descartes' letter, Leibniz (who was well acquainted with Dalgarno's and Wilkins' works) rebuffed the Cartesian criticism: "It is true that this language depends on the true philosophy, but it does not depend on its perfection. That is to say: this language will be constructed despite a still not perfect philosophy."[18]

IV. Rhetorical Patterns for the Logic of Scientific Knowledge

To search, to find, to judge, to record, to transmit—who today could conceive these activities as the interconnected parts of a single discipline? Francis Bacon, between 1605 and 1623, understood the logic in a manner far different from that of a professor of logic of our time. He assumed the five above mentioned terms and, so to say, boxed each inside the others. He mixed up portions of the logical or "dialectical" tradition with portions of the rhetorical tradition. In the *Advancement of Learning*, logic is divided into four parts according to its different functions and the different objectives to which man tends: man finds what he is searching for; he judges what he has found; he records what he has judged; he communicates what he has recorded. The logic therefore comprehends four arts: the art of inquiry or *invention*; the art of examination or *judgment*; the art of custody or *memory*; the art of elocution or *tradition*.[19]

The method of scientific research of the *Novum Organum* cannot be identified (as many epistemologists often do) with Bacon's logic. The interpretation of nature is only one of the two parts that make up the art of invention (the other is "literary practice"). The art of invention, in turn, is only one of the four subdivisions of Bacon's logic.

Bacon deals with problems concerning the understanding of nature in typically rhetorical terms of a discussion of invention of arguments. The art of questioning in a discussion serves Bacon as an illustration for the art of questioning nature. My intention is to show here how much Bacon's discussion on scientific method owed to the dialectico-rethorical tradition and how a number of his theories about that method were transplanted from the field of rhetoric. Bacon's theses maybe summarized as follows:

1) Traditional philosophy has not failed in what it attempted. It is quite capable of preserving and transmitting sciences and inventing arguments to outwit others in a discussion:

> In sciences founded on opinions and dogmas, the use of anticipations and logic is good; for in them the object is to command assent to the proposition, not to master the things.[20]

We are not interested, says Bacon, in popular controversial arts, and the new logic does not pretend to serve the ends of traditional logic:

> The logic which is received, though it be very properly applied to civil business and to those arts which rest in discourse and opinion, is not nearly subtle enough to deal with nature.[21]

2) The art of memory operates on two different levels: that of "old popular sciences" and that of a "completely new" scientific method of natural enquiry.[22]

3) By including the "helps for memory" in the new logic, certain concepts pertaining to traditional rhetoric became incorporated in the "interpretation of nature."

4) Bacon stresses on many occasions the discrepancy between the aims and methods of ordinary logic and of scientific logic. But where that section of a new

logic—the aids to memory—are concerned, this does not stop him from adopting a mode of reasoning almost identical to that he had employed for the "art of discourse" or "ordinary logic." In the case of discourse, a multitude of terms and arguments had to be recollected and organized; in the case of scientific method, this applies to the multitude of instances. If we wish to utter coherent persuasive discourse and invent arguments, we must: dispose of an extensive collection of argument (*promptuaria*), and possess the rules for restricting a boundless field, reducing it to the proportions of a specific discourse (*topica*). The art of memory permits the realization of both these requirements.

For Bacon this procedure undergoes little change when applied to scientific knowledge:

> The aids to memory (*Ministratio ad memoriam*) fulfill the following mission: from the confusion of particular instances and the bulk of natural history of particular history is selected, and its elements are disposed in an order such as to enable the mind, according to its own capacity, to work thereon. . . . Firstly we will exhibit which are the things to be looked for about a specific problem, and that is like a Topic. Secondly we will exhibit the order in which they must be disposed and distributed in the Tables. . . . The aids to memory therefore include the invention of loci and the method for the construction of tables.[23]

Bacon is no less explicit in the *New Organon*, II, 10:

> The interpretation of nature . . . is divided into three ministrations: a ministration to the senses, a ministration to the memory, and a ministration to the mind or reason. For first of all we must prepare a Natural and Experimental History. . . . But natural and experimental history is so various and diffuse, that it confounds and distracts the understanding, unless it be ranged and presented to view in a suitable order. We must therefore form the Tables of Arrangement of Instances, in such a method and order that the understanding may be able to deal with them.[24]

Many theorists of rhetoric in the Renaissance had stressed the importance of loci as a means of restricting an otherwise boundless field and of classifying the materials. According to Melanchthon, to pick an example at random:

> Loci . . . advise us when material is to be sought or generally as to what should be selected from the great heap available and in what order it should be classified. For the loci of invention both in the writings of dialecticians and in the orators do not lead us to the discovery of material as to the solution of the problem of choice.[25]

5) Bacon, as I pointed out in my book on him, used the rules theorized by Petrus Ramus for defining forms. The well-known *tables* of Baconian induction are defined as "aids to memory." Bacon's interpretation of nature adapts fifteenth-century rhetorical and philosophical mnemonics to its own ends. Typically rhetorical concepts were thus transplanted by Bacon into the scientific field of natural research. He devised the tables or instruments of classification to organize reality and thus enable the memory to assist intellectual operations. For Bacon this vision of a form of logic assisting in the classification of instances collected in a great natural history was particularly seductive.

I think that Bacon's definition of his method as a "thread" guiding mankind through the "chaotic forest" and "complex labyrinth" of nature is very important. Bacon's doctrine of scientific knowledge is entirely conditioned by his concept of the universe as a "forest" filled with "so many ambiguities of way," "deceitful resemblances of objects and signs," "natures irregular in their lines so knotted and entangled."[26] One of the method's first objectives is to set order in the variety of nature. In the *Partis secundae delineatio*, Bacon admits that truth emerges more readily from error than from confusion, and reason finds less difficulty in modifying incorrect classifications than in delving into disorder.[27]

V. Rhetoric and Method of Science

As Marie B. Hesse pointed out, "it was a favourite pastime in the nineteenth century to criticize Bacon for not being a Galileo or a Newton."[28] During the 1950s and 1960s, that pastime was not abandoned. Two different negative appraisals converged on Bacon's philosophy. According to some neopositivists and Popperian epistemologists, Francis Bacon was a model or a champion of what science has never been and never will be: a kind of knowledge deriving from observations, a process of accumulation of data, an illusory attempt to free the human mind from presuppositions. According to the representatives of the Frankfort School, Francis Bacon was precisely the opposite: the symbol of what science has been up to now and should no longer be—the impious will to dominate nature and mankind.

According to the philosophers of our century who exalted scientific knowledge, Bacon has nothing to do with science. According to the philosophers who criticized scientific knowledge or accused it of many sins, Bacon was the very "essence" of science. Although at variance over every philosophical problem, the two philosophical (and influential) parties agreed on rejecting, for opposite reasons, Bacon's philosophy.

Only by avoiding "the plumb and the weight" of the texts and "flying to the most general conclusions" on the basis of some philosophical handbook, it is possible to attain the results given as an example, in the two following quotations:

The inductivist logic of discovery (wrote Imre Lakatos) is the baconian doctrine according to which a discovery is scientific only if it is guided by facts and not misguided by theory.[29]

I read in the *New Organon*:

The manner of making experiments which men now use is blind and stupid. And therefore, wandering and straying as they do with no settled course, and taking counsel only from things as they fall out, they fetch a wide circuit and meet with many matters, but make little progress.[30]

According to Horkheimer and Adorno, in the *Dialektik der Aufklärung*:

The infertile happiness of knowledge is lascivious according to Bacon as according to Luther. Not that kind of satisfaction that men call truth is important, but only the operation, the successful procedure.[31]

But I read in the *Advancement* and in the *New Organon*:

The pleasure and delight of knowledge and learning is far surpasseth all other in nature . . . So we must from experience of every kind first endeavour to discover true causes and axioms; and seek for experiment of Light, not for experiment of Fruit.[32]

In the 1960s every connection of Bacon's theory of scientific method with the heritage of the rhetorical tradition was regarded as confirming the absolute irrelevance of his theory. Science was seen as having nothing to do with Francis Bacon and with rhetoric. Science was the realm of theories and of experiments confirming or disproving theories. The study of rhetoric and the art of memory was conceived as an innocuous, purely academic investigation about strange, obsolete, and totally irrelevant techniques.

In the 1970s, the image of Bacon as the "false start" (according to Alexandre Koyré's expression) of modern science gave way to that of Bacon as the "transformer of the hermetic dream." I dedicated a large part of my book on Bacon (first published in Italian in 1957) to illustrating his indebtedness to the magico-alchemical tradition. After the publication of F. A. Yates' important book on Bruno (1964), Bacon was seen as an exponent, in more modern idiom, of the ideas and values of the hermetic tradition. Bacon, who was extremely fond of classifying and typifying, saw in magic a typical manifestation of "phantastical learning," in scholastic disputations a kind of "contentious learning," and in Ciceronian humanism the expression of "delicate learning." That he was variously conditioned by these three cultural forms should not blind us to the fact that he attempted to formulate a new image of science in violent contrast with magic, scholastic thought, and the tradition of Italian and English humanism.[33]

The situation has today entirely changed. But I suspect that the recognition of a whole series of problems and difficulties has been transformed in the last decades into a string of impossibilities: since experiment no longer plays the decisive function attributed to it by inductivism, then science contains only theories; since a linear and continuous development is absent from science, only a series of "choices" between theories remains; since an attempt at a reconstruction on rational lines stumbles on elements that cannot be reconstructed, then one is left with appeals to individual and collective psychology; lastly, since we plainly see the connections between scientific method and rhetoric, then we have only rhetoric, and science no longer exists.

The history of mankind is full of myths, of religious and metaphysical ideas, of conjectures, and of modes of perceiving the world that will never be integrated into science. In other cases, however, some of these myths and conjectures influence science and become an integral and component part of science. They "dissolve" into science. I think that baconian rhetoric was actually dissolving into the "baconian sciences" of the seventeenth and eighteenth centuries. And I think also that Thomas Kuhn is perfectly right in distinguishing, in relation to the first centuries of our modern era, baconian from classical sciences. But the right thing to do is perhaps to retain a dose of pessimism. Francis Bacon was also right: it is impossible to root out all the idols from men's minds.

NOTES

1. J. Pitton de Tournefort, *Eléments de botanique ou méthode pour connaître les plantes* (1694), vol 6 (Lyon, 1797), I, p. 45.

2. J. Pitton de Tournefort, *Institutiones rei herbariae* (Parisii, 1700), pp. 12, 13, 15.

3. C. Linnaeus, *Systema naturae* (Holmiae, 1758), p. 11.

4. J. Wilkins, *An Essay towards a Real Character and a Philosophical Language* (London, 1668), p. 289. The same is the design of the tables in the work of G. Dalgarno, *Ars signorum vulgo character universalis et lingua philosophica* (Londini, 1661). Cf. *The Works of George Dalgarno of Aberdeen, presented to the Maitland Club* (Edinburgh, 1834).

5. J. Wilkins, *Essay*, p. 21.

6. G. Camillo, *Idea del Theatro* (Firenze, 1550); *Opere* (Venezia, 1584). See F. Secret, "Le Théatre du Monde de G. C. Camillo et son influence," *Rivista Critica di Storia della Filosofia* 11 (1959), pp. 418–436.

7. Kinner's passage is in a letter to Samuel Hartlib that has been published by B. De Mott, "The Source and Development of John Wilkins' Philosophical Language," *Journal of English and German Philology* 57 (1958), pp. 1–12. See also, "Comenius

and the Real Character in England," *PMLA* (1955), pp. 1068–1081; "Science versus Mnemonics," *Isis* 48 (1957), pp. 3–12.

8. B. de Fontenelle, *Eloge de Tournefort. Histoire de l'Académie des Sciences* (Paris, 1708), p. 147.

9. Paolo Rossi, *Clavis Universalis: arti della memoria e logica combinatoria da Lullo a Leibniz* (Milano–Napoli: Ricciardi, 1960); new edition (Bologna: Il Mulino, 1983, 1988). See also, "The Legacy of Ramon Lull in Sixteenth-century Thought," *Mediaeval and Renaissance Studies* V (1961), pp. 182–213; "Che cosa abbiamo dimenticato sulla memoria?" *Intersezioni* 7 (1987), pp. 419–428; "The Twisted Roots of Leibniz' Characteristic," in *The Leibniz Renaissance*, M. Mugnai, ed. (Firenze: Olschki, 1989), pp. 271–289; "La memoria, le immagini, l' enciclopedia," in *La memoria del sapere*, Pietro Rossi, ed. (Bari: Laterza, 1989), pp. 211–238; "Il paradigma della riemergenza del passato," *Rivista diFilosofia* 80 (1989), pp. 371–392; "Le arti della memoria," *Mondo Operaio* 12 (1989), pp. 103–109.

10. M. Adanson, *Familles des plantes* (Paris, 1763), p. xcv.

11. "Necessitas hitoriae naturalis Rossiae, quam Praeside C. Linnaeo, proposuit Alexander de Karamyschew, Upsaliae, 1766, maij, 15," in Ch. Linné, *Amoenitates Academicae* (Holmiae, 1769), VII, p. 439.

12. See H. Daudin, *Les méthodes de classification et l' idée de série en botanique et en zoologie de Linné a Lamarck* (Paris, 1926); A. Arber, *Herbals: Their Origin and Evolution* (Cambridge, 1953); F. Dagognet, *Le catalogue de la vie* (Paris: Presses Universitaires de France, 1970).

13. C. Linnaeus, *Systema Naturae* (Holmiae, 1758), II, p. 874.

14. "Reformatio Botanices quam, Praeside D. D. Car. Linnaeo, proposuit Johannes Reftelius, Upsaliae, 1762 decembr. 18," in Ch. Linné, *Amoenitates Academicae* (Holmiae, 1769), VII, p. 306.

15. See T. and J. Carraeras y Artau, *Filosofia cristiana de los siglos XIII al XIV* (Madrid, 1939–1943), 2 vols.; G. N. Hilgarth, *Ramon Lull and Lullism in Fourteenth-century France* (Oxford, 1971).

16. See F. A. Yates, *Giordano Bruno and the Hermetic Tradition* (London: Routledge and Kegan Paul, 1964); *The Art of Memory* (London: Routledge and Kegan Paul, 1966).

17. See Paolo Rossi, *Francis Bacon from Magic to Science* (Chicago: University of Chicago Press, 1968), pp. 166–172.

18. R. Descartes, *Oeuvres*, Ch. Adam et P. Tannery, eds. (Paris: Cerf, 1897–1910), 12 vols., II, pp. 76, 82; L. Couturat, *Opuscules et fragments inédits de Leibniz* (Paris, 1903), pp. 27–28.

19. F. Bacon, *The Works of Francis Bacon*, R. L. Ellis, J. Spedding, and D. D. Heath, eds. (London, 1887–1892), vol. 7, III, pp. 383–384; I, pp. 615–616.

20. *Novum Organum*, I, 29.

21. *The Works of Fr. Bacon*, IV, p. 17.
22. *The Works of Fr. Bacon*, IV, p. 435.
23. *The Works of Fr. Bacon*, III, pp. 352–353.
24. *Novum Organum*, II, 10.
25. P. Melanchton, *Rhetorices Elementa* (Venetiis, 1534), p. 8.
26. *The Works of Fr. Bacon*, IV, p. 18.
27. *The Works of Fr. Bacon*, III, p. 553.
28. M. B. Hesse, "Francis Bacon's Philosophy of Science," in *Essential Articles for the Study of Francis Bacon*, B. Vickers, ed. (London, 1978), pp. 114–139.
29. I. Lakatos, "Popper on Demarcation and Induction," in *The Philosophy of Karl Popper*, P. A. Schilpp, ed. (La Salle, Ill., 1974), p. 259.
30. *The Works of Fr. Bacon*, IV, p. 70.
31. M. Horkheimer and Th. W. Adorno, *Dialektik der Aufklärung. Philosophische Fragmente* (Amsterdam, 1947), p. 15.
32. *The Works of Fr. Bacon*, III, p. 317; IV, p. 71.
33. See Paolo Rossi, "Hermeticism, Rationality and the Scientific Revolution," in *Reason, Experiment and Mysticism in the Scientific Revolution*, M. L. Righini Bonelli and W. R. Shea, eds. (New York: Science History Publications, 1975), pp. 247–273.

On Deciding What to Believe and How to Talk about Nature

DUDLEY SHAPERE

I The Language of Science

A recent article by Lada and Shu states one of the central problems of modern science in a particularly lucid and philosophically instructive way. According to them,

> Modern star-formation research has as an objective the elucidation of the physical process by which a giant molecular cloud transforms a small fraction of its mass into numerous self-gravitating balls of gas that have just the right range of masses—roughly, 10^{-1} to 10^2 times the mass of the sun ($1M_{solar} = 2 \times 10^{33}$ g)—to fuse the primary product of the Big Bang, hydrogen, into heavier elements by way of nuclear reactions.[1]

Though this particular quotation is a statement of a problem or objective, it is also the case, as can be seen in other parts of the Lada-Shu paper, that directions in which to go in trying to answer the problem, and proposals of answers to the problem, are phrased in language that is similar in important respects. It is what many philosophers and historians of science might call a final-form statement, the product of a great deal of prior thinking, observing, and experimenting, presented in a form considered appropriate for publication in a professional journal. The form of language used here is typical in important respects of that used in sophisticated publications and conference reports in many areas of modern science. What these respects are need to be specified; we are not helped much by calling it "technical vocabulary," as though that mere expression explained its origin and functions. But however it is to be characterized, this sort of thing is the language of polished science. It is a language in which problems, lines of research, and explanatory solutions that have been worked over, refined and polished are expressed. It is language in which, in a sense that also must be clarified, scientists believe their problems, investigations, and solutions *should* be expressed. Thus three problems arise with regard to this language. First, how do scientists come to formulate their problems, lines of research, and theoretical explanations in such ways? Second, how are we to give an account of what those "ways" are that will illuminate the processes and aims of science? And third, to the extent that use of this language is normative, what is the basis of that normative force?

In this paper, I will try to suggest answers to these three questions. The approach taken is designed to bring out certain points in the final section concerning a fourth question: for what reasons, to what extent, and in what ways must such language be supplemented by other devices in order to convince or convert others to a new theory, research problem, or point of view in science?

II The Development of Scientific Rules and Language

Let us take a closer look at the Lada-Shu statement of the problem of "modern star-formation research." Here a program of research is outlined, and the terms in which the program is expressed clearly depend on prior conclusions. In particular, stars are declared to be self-gravitating balls of gas, and to have a definite mass-range related to the fact that they are capable of producing nuclear fusion reactions. The entire program is put into the context of Big Bang cosmology. Further, the program is fleshed out (in passages that I have not quoted) in terms of a number of specific problems that specify ways in which our understanding of the process of stellar birth is incomplete, and which must be answered if that process is to be understood. A range of possible answers to those problems is explicitly stated for consideration, and available evidence about the alternative possibilities is discussed, and means of gaining further evidence outlined.

The program is thus formulated within, and in terms of, a background framework of ideas that are taken for granted. That stars are balls of gas was established in the 1920s, overthrowing theories like that put forth by James Jeans and others, according to which they are solid objects. That they form by collapse of gas clouds came as a gradual realization stemming from a succession of developments including the following: the recognition that the dark patches crossing the Milky Way are not regions devoid of stars, but are due to the presence of obscuring matter; studies of the mechanics of large clouds of gas (and dust), and in particular the dynamics of their collapse and fragmentation; the discovery, arising in turn from the theories of stellar evolution developed since the late 1930s, that young hot stars inhabit gas clouds; and the resulting rejection of theories according to which stars originate by emission of matter from spatial singularities, as had been proposed independently by Jeans, Jordan, and Ambartsumian. That stars are composed almost entirely of hydrogen (along with helium comprising about 98 percent in the case of the sun) was found by a painstaking analysis, by Russell and Payne, of features of spectra that appeared superficially to suggest that the contrary is the case. That the hydrogen content of stars provides the fuel for the energy of stars through most of their lifetimes (and also what happens to stars when the hydrogen available as fuel is used up) was a major dividend of the development of a sound theory of nuclear reactions in the early 1930s. That the clouds from which stars are born are molecular and very large is a result that has come to be accepted only in the past decade, on the basis of a variety of considerations including satellite observations in regions of the electromagnetic spectrum outside the optical range. That the masses of the resulting stars would have to lie within a certain range was a consequence not only of the fact that observations failed to reveal any shining stars outside that range, but also (as regards the lower limit of the mass range) of the requirement that the weight of overlying material would have to be sufficient to heat the stellar interiors to the temperatures required for the relevant nuclear reactions to take place.

As we see from the quotation, this variegated body of ideas formed a background of ideas in terms of which the question of the birth of stars could be formulated, and in terms of which research on that question could be guided. In turn, though in some cases these background ideas were, and remain today, incomplete in certain specifiable respects, they were accepted for employment in such formulation and guidance because, among other things, they were highly successful in accounting for the various observations relevant to them, and because what rival accounts had been proposed had been ruled out.

Those background ideas in turn had their backgrounds, in terms of which they themselves and the problems, theses, and research concerning them had been formulated. I referred above to the theory of stellar evolution as part of the background of the problem of stellar birth. Over a period of half a century since the new nuclear physics had been applied to the problem of the source of stellar energy (Why do stars shine?), astrophysicists have obtained a deep and detailed understanding of the colors and chemical makeup of stars, the relations among their masses, luminosity, and lifetimes, the reasons for their distribution on a graph plotting luminosity against color (the Hertzsprung-Russell diagram, with its "Main Sequence" diagonal, giant and supergiant branches, and white dwarf region), and the later stages of evolution after available hydrogen has been converted to helium.

Those were the kinds of things a successful theory of stellar evolution was supposed to account for, the domain of that area of investigation. Too, what a theory of stellar evolution would have to be like to be successful was itself a product of inquiry guided by still prior background ideas, and was formulated in their terms: the theory had to be formulated in terms of such factors as nuclear reactions and their rates, theories of stellar structure, theories of radiative and convective transfer of energy, and so forth. Earlier theories of stellar energy production—as effects of cutting out hearts of sacrificial victims, or as produced by constant meteoritic impacts, or by gravitational collapse—and, where the question was even raised, of stellar evolution, were no longer in scientific contention as shaping the character of possible solutions of the problem. Similarly, much understanding of the death of stars, and the way in which the fusion-generated higher elements are dispersed into the interstellar medium, has also been obtained, though at this end of stellar lifetimes as at the beginning, much remains to be understood.

Thus there is a background to the background in terms of which the problem of stellar formation could be expressed in the Lada-Shu manner. But there is yet a deeper historical background of that background to the background! We can strip away both the immediate and the indirect sophisticated backgrounds that govern the Lada-Shu formulation—the theories of stellar evolution, nuclear physics, and so forth—and merely ask a rather general and everyday sort of question: How are stars born? When we do this, we find that that simple question, askable about so many things in everyday experience, was not the sort of question that was always considered askable about stars. Aside from some primitive myths, for most of history, stars were taken as being fixed and eternal, or at best created all at once "in the beginning." Thus there had to arise, somewhere along the line, some background of ideas that made that general question itself one that could and should be asked about stars. (The same is true for the question: How do stars evolve?) This background was indeed furnished, though only gradually. A key step came when Laplace and Kant applied Newtonian gravitation to the analysis of pinwheel-shaped objects ("spiral nebulae," as they were called until fairly recently) to suggest, on what we now realize were quite perilous analogies with everyday whirling objects and fluids, the first theories of the birth of stars and planetary systems. Though such "nebular hypotheses" and their later formulations were stated with enough clarity to allow their defects ultimately to be noted and alternatives suggested, it was only with the modern theory of stellar evolution that, as will become evident shortly, the problem of stellar birth could be stated with sufficient precision to permit details of the process to be proposed and investigated.

In addition to a deep historical background, there has also developed a much broader context in terms of which specific questions, research, and theories are to be formulated. Increasingly during the period 1964–1980, this broader background has been taken to be the Big-Bang theory, now fused with the immensely successful Standard Model of elementary particles and forces. Among the areas of investigation that must now be framed within this background context are the origin and evolution of the chemical elements; the origin and evolution of galaxies; our present area, the origin and evolution of stars and their companion objects; and even, as I will suggest later, the origin and evolution of life. Thus, for example, the collapse theory of stellar births has now come to be seen as a natural outgrowth of the view that the expanding primordial hydrogen and helium could fragment into clouds that, under the right conditions, would collapse under gravitational attraction to form galaxies and clusters thereof, and that the fragments of gas would fragment still further to form smaller objects.

Within the context of that background, the question of stellar birth can now be approached with the considerable precision of formulation we see in the Lada-Shu paper. We know, in far more detail than was possible for Laplace and Kant, what needs to be accounted for. We know that distinctions may be necessary between a theory of the origin of small-mass stars like the sun and very massive stars; here, as in the Lada-Shu paper, I consider only the former. Specifically, in terms of a large body of successful background ideas—where we have learned also what it is for those ideas to be "successful," and also other criteria that must be met—we know what has to be produced, where it will be produced, and a good deal about the circumstances under which it will be produced. Laplace and Kant knew only of hazy pinwheels in space, which they visualized as contracting under Newtonian gravitation to form stars; they had no other information about the pinwheels to go on, such as their sizes, distances, composition, rotation rates, temperatures, or, obviously, the presence and strengths of magnetic fields or spectral characteristics. Theories of radiative and convective transfer, turbulent flow, knowledge about weak electromagnetic fields and the behavior of particles in their presence, and, of course, the theory of spectral line formation are now just a few of the multifarious ingredients added to the background to be applied as relevant to the problem. Further beyond Laplace and Kant (or even Jeffries and Jeans and Chamberlain and Moulton), we also have the means of obtaining the information required to construct and test specific theories, means that those earlier thinkers could not even have thought of. For instance, since stars are born within "cocoons" (or "placentas") of gas and (in today's adolescent galaxy) dust, light in the visual spectrum is absorbed, and the birth process is hidden from classical means of observation. But (another piece of background information) infrared light will escape from the cloud, carrying information about what is going on inside, information that we have the means to decode. The relevant infrared spectral window, in the submillimeter and millimeter wavelength range, was opened only in the 1970s: new radiotelescope capabilities and infrared satellite observatories are the means of access. Finally, as our quotation and supplements thereto show, the specific problems that need to be dealt with are also clearly statable.

III The Character and Roots of Background Ideas

From this brief survey, we get a picture of scientific change in which, over the development of science, certain ideas become accepted to a degree such that they can guide the formulation and investigation of problems and the description of the objects and events to be investigated. The language in terms of which those problems, descriptions, and research programs are formulated is shaped or interpreted in the light of those accepted background ideas. In cases of which the reformation of chemical nomenclature by Lavoisier and his associates is the paradigm and extreme example, the language is explicitly reshaped; in other cases, like that of the term 'star' when it becomes necessary to distinguish the birth-processes of small from massive stars, or when it becomes similarly necessary to distinguish the origin and nature of island arcs, Hawaiian-type mantle-plume products, and Seychelle-type continental fragments, the older language is merely reinterpreted. In this sort of process, alternative views, ranging from mythical celestial transports of heroes and queens to white-hole speculations made possible by time-reversals in black-hole theory, are rejected, some as being scientifically possible but false, others as being nonscientific in senses that are clearly specifiable in individual cases. In the case of Orion and Andromeda, for instance, there is, among the established body of background beliefs employed in science, no ground whatever to suppose it possible for people to be transferred to the sky and converted into constellations. Thus, a distinction emerges—has emerged, gradually and with backsliding, to be sure—between considerations that are to count as internal to science, and ones that are to count as external. In the account I have given, this is not a distinction based on some *a priori* criterion of demarcation of the scientific from the nonscientific, some criterion that transcends the processes of science themselves. Nor is it always a clean and clear distinction, though in some areas of science, and some topics within areas, it is quite sharp and clear. The body of background beliefs is a heterogeneous one, not a unified perspective, and it is also not a vague "point of view" that is applied indiscriminately in all scientific inquiries. It consists of a variegated multitude of ideas that have, at least *prima facie*, been found to be successful in accounting for bodies of putative information. The background ideas are not even necessarily fully formulated, nor even necessarily consistent with one another. Nor are they necessarily free of all problems, even of all known ones, though we have come, through the process of inquiry, to insist that they meet certain conditions of freedom from objection. They leave open many alternative possible problems, possible lines of research, and possible theories, even though they exclude others.

The distinction that thus emerges is not confined to what have traditionally been categorized as substantive beliefs. In other work, I have argued that the same sorts of changes enter into determining what is to count as "observational"—how that term, which once was treated as "metascientific," is to be understood. [2] In its case, new background information provides bases for new kinds of access to information, where that information is specified by that background. Again, what it is to be an "account" or "explanation" is something that has evolved historically in ways parallel to my example: whereas earlier accounts of what it is to be an account in physics insisted that every observed feature of a system be deducible from initial conditions according to laws, in today's quantum world accounts are not of that sort, and such accounts can still be considered legitimately acceptable, even though they would not have been so considered in earlier days. The characteristics that are to be demanded in an acceptable theory are shaped by background beliefs, and those background beliefs can change in the light of new findings.

When the background beliefs satisfy the constraints that have arisen in the course of inquiry, they can become normative rules governing the conduct of research, the expected forms of answers to problems, and the ways of formulating scientific problems: there are good reasons, in the background of modern science, why physical theories of elementary particles and forces should be formulated as quantum theories, and still more particularly as quantum field theories, and still more particularly as local gauge theories. Again, the rules are not "metascientific" ones, established independently of inquiry, but are arrived at in the course of inquiry. They are not formal, logical, but depend on the content of accepted beliefs. Thus dependent, they are themselves subject to question, alteration, rejection, or replacement in the light of inquiry.

The existence and employment of background ideas in the formulation of problems, lines of research, and theoretical possibilities, and the status of those background ideas as normative principles, does not imply that they themselves cannot be questioned, modified, or rejected in the light of the results of inquiry. The possibility of criticizing them arises from the fact, missed in so many post-positivistic views of science, that there is after all something that is "given" in observation, something that is independent of the background that serves as reasons in the conception, conduct, and interpretation of scientific research. True, it is not the given as conceived by earlier empiricism and positivism, a "given" in pure perception. Rather, it is a given in the sense that, (a) having been marked out as significant by our best available background ideas, (b) having been appropriately described in terms of those background ideas, and (c) having been made accessible by application of background ideas, the specific character or value we find it to have is independent of—not determined by—those background ideas. Though the concept of nuclear magnetic moment is highly dependent on background ideas for the selection, characterization, and means of determination of the specific property so described, the value found by those means of determination may or may not agree with the theories formulated in terms of that background information, and in some specifiable sorts of circumstances, disagreement of that "given" value can combine with other pieces of the variegated available background to lead to rejection of any particular piece of background, and even of rather large portions thereof.

Thus, on the present view, there are rules that guide the formulation of scientific problems, research, and alternatives, and those rules can attain a normative status. But in keeping with criticisms that led to the rejection of the positivistic and other traditions in the philosophy of science, there is no certainty or necessity, no *a priori* access, to those rules. They are open, in ways that can be specified, to criticism, rejection, and replacement. Nor are they necessarily universal, applying to all science: whether there are any such universal rules, even for contemporary science, even for any one area of contemporary science, is a contingent matter. I have mentioned the highly successful Standard Model of elementary particles and the normative status of local gauge field-theoretic formulations of fundamental physical theories that results from that success. But theories in biology, or even in chemistry or stellar or galactic astronomy, are not required to be formulated as gauge theories. The rules we employ in science are, as a matter of contingent fact, local, and are learned, arrived at, and therefore are themselves contingent. They are set by what we have learned, or at least have reason to believe have been learned, in a sense of "reason" that we have also learned.

This picture of science looks backward from the perspective of present beliefs, problems, and rules, to the background in terms of which those aspects of present science have been constructed. This backward-looking, however, is not "Whiggish": understanding of the way background-framework ideas develop must involve understanding of the background of past ideas, including the alternatives available at past times in the light of the background prevailing then, even though those alternatives may have been rejected. For it is crucial to the present view that the distinction between what is "scientific" and what is "nonscientific" has evolved: what was or was not scientifically relevant for a Galileo or a Newton may have included things that would later be rejected as irrelevant, as nonscientific; or it may simply not have been clear, at some given stage, what was relevant and what was not. But if we are to gain a full understanding of the enterprise of science, the evolution of that distinction, and its present existence, must not be ignored. (This is why one must be very careful in drawing conclusions about *science* from studies of single cases, especially ones taken from earlier epochs, when the line between science and nonscience, if drawn at all, may have been drawn quite differently from the way it was drawn later.)

But if our ways of inquiring and talking in science are at every step shaped by background beliefs, where did it all begin? Was there some original set of background ideas, and where did our ancestors get them? As with all inquiry, there is a background to this question and its possible answers too. I take that background to consist of what we have learned about the origins of philosophical and mythical ideas, together with what we have learned about the place of the human species in the evolutionary process. Classical scholars have shown how the vocabulary of early Greek philosophy arose from (as Eric Havelock has put it[3]) a "stretching" of commonplace everyday terms: terms like *kosmos*, originally having to do with arrangements of parts of an everyday thing; of *harmonia*, originally a fitting together of those parts; of the idea of plenum, whose cognates are used by Homer only in reference to people's bellies after a full meal. Viewed from such Greek "stretchings" into the future, refinement and extension of such language can then be seen in the subsequent development not only of philosophical vocabulary, but also as an integral part of the development of science, particularly in such scientific activities as the development of successive nomenclatures of types of material substances, from Theophrastus through Lavoisier to the expression of properties of elementary particles as quantum numbers, and in the long process eventuating in the Lada-Shu formulation of the problems of star formation.

As I said, the other piece of "background" which must shape our understanding of the roots of inquiry—the question, Where did it all begin?—consists of the theory of biological, and in particular of human, evolution. In the light of that framework, it is only natural to expect the origins of Greek philosophical vocabulary to be found in the everyday and commonplace, for those were the original and pressing occupations of earlier human beings; where else would one look for the sources of scientific and philosophical ideas, unless one succumbs to the fantasy of supposing that the human mind has special access to the rules in terms of which nature is to be understood and talked about? From this perspective, the very starting-point of our inquiries about the sources of our ideas, including what we take to be justificatory reasons, must be in the everyday life of our earlier ancestors, and thus ultimately in the history of life as a whole. This is the framework in terms of which our questions about the nature of science and its results must be expressed, in the same sense that there is a background in terms of which the problem of stellar formation is expressed by Lada and Shu. The history of human thought about the world, and about what that thought should be like and how it should be expressed in language, is a history of departures from everyday experience and the language and criteria suggested by that experience in the attempt to come to terms with it and by antecedent such efforts in the history of life. That is the only approach to understanding science and its language that is consistent with what we have learned from studies of human evolution and of the development of ways of talking and thinking about the world. And that is why, in all their various guises, searches for *a priori* and necessary truths, which are allegedly accessible to the human mind independently of inquiry as we have learned to conduct it, constitute one of the last remaining sanctuaries of superstition.

IV *Rules, Reasons, and Rhetoric in Science*

But though philosophies of science that held that science is based on rules that are necessary, absolutely universal, and discoverable *a priori* are wrong, it is equally mistaken to suppose that because there are no universal and necessary rules, there are therefore no rules at all, or that without universal and necessary rules there is no justification for any scientific conclusions, or that without such absolute rules of science a distinction between the scientific ("internal") and the nonscientific ("external") is illegitimate. The view outlined in this paper steers a clear course between the Scylla of absolutism and the Charybdis of relativism, and is framed in compatibility with our best conclusions about the origins and development of human thought. It shows how, as we have gone beyond the humble transactions of everyday experience, the terms in which we think and talk about nature are altered, and how rules can develop—have developed—that distinguish the scientific from the nonscientific, and that can develop a normative status with regard to inquiry.

This viewpoint ties the development of the language in which scientific problems, research, and conclusions are expressed to the best conclusions at which science has previously arrived. To put the point in another way, the language of science, or at least its interpretation, is *framed*—has come to be framed—so as to reflect, or presuppose, a body of conclusions that there is best reason to accept and least reason to doubt. But also, as we have seen, at least some pieces of background information attain a normative status, and there is the implication that scientific problems, research, and results in specific areas *ought* to be expressed in the language indicated by the background framework of that area of inquiry. It is so formulated, as we have seen, by Lada and Shu for their particular subject. And further, there is the implication that people *ought* to be convinced or converted by the account presented, with its evidence and reasoning, expressed in these ways. Two of the definitions of the term "rhetoric" given in the Oxford English Dictionary are: "The art of using language so as to persuade or influence others; . . . Speech or writing expressed in terms calculated to persuade. . . ." In those senses, the "final-form" language of science *is* the rhetoric of science: it is a form that is calculated to persuade, at least in the case of those who understand the language (*i.e.*, know its background), precisely because it is supposed to incorporate the scientific considerations giving rise to the problem, research, or conclusions so expressed; to embody, by its very formulation, the reasons and evidence that ought to convince or convert. Naturally, "ought" does not imply "will," and as Aristotle pointed out, it may be necessary to supplement that language, in the light of an understanding of human character and emotions, "to know their causes and the way in which they are excited."[4] And even then, as the Philosopher understood, "there are people whom one cannot instruct,"[5] so that quite different means of persuasion, perhaps ones that are in specifiable ways extrascientific, may be required.

Yet the proposal that something called rhetoric has a role in science is today generally taken to involve more than this. For somehow the "final-form" language of science, with all its embedded reasoning and evidence, is in many circles considered to be intrinsically inadequate, and even as irrelevant as a means of convincing, of converting scientists and others to a point of view. We do not have to look far for the grounds of this viewpoint: they stem from the arguments that led to the rejection of positivism. Let me survey some of the conclusions of those arguments.

In the letter of invitation to this conference, Professor Pera explained its motivations in an admirably clear manner:

> The conference stems from a challenge issued by the "new philosophy of science": If there are no universal and precise methodological rules, how do scientists, during a theory-change, come to convince or convert their community to a new theory or way of seeing the world?[6]

This is a fair statement of much current thinking on a central issue in current philosophy of science, posed by the rejection of positivism. However, there remains an important ambiguity in its counsel to examine the role of rhetoric in science: Does the absence of universal and precise methodological rules require examination of linguistic techniques and arts beyond the "final-form" language of science, techniques and arts that must be superimposed on that language or supplement it, or even, perhaps, replace it? On the one hand, I have argued that the absence of universal and precise methodological rules does *not* raise any difficulty about the language of science: we can understand the character and roots of scientific language as distinguished from other methods of persuasion. From this point of view, the study of the role of rhetoric in science encompasses the final-form language of science itself; and that study can be supplemented by examination of whatever uses of language might help convince others of a scientific claim, even though those supplemental uses may themselves be "external" and, as the OED also says, "artificial," at least from an internal scientific perspective.

But on the other hand, there is a more extreme view of rhetoric and its role in understanding science. It also stems from arguments involved in the rejection of positivism, but on ones other than the rejection of universal and precise methodological rules. Here is where we find the contention that all the supposed reasoning and evidence lying behind the final-form statement of scientific ideas and approaches are intrinsically inadequate to establish the idea or approach in question. Two arguments to this effect stand out in the literature: one, that scientific evidence (and reasoning) *underdetermine* the particular scientific conclusion they are supposed to support or establish; the other, that for any particular body of evidence or reasons, there are always a large and in principle infinite number of possible alternative explanatory accounts. There is thus said to be a gap between scientific reasoning on the one hand, and scientific conclusion (statement of problem, line of research, or solution) on the other. That gap, so the story goes, can only be filled by considerations that go beyond the purely scientific evidence and reasons—for example, by rhetorical techniques, where "rhetoric" now consists of wholly extrascientific means of persuasion. Even stronger theses are often held to emerge from such arguments for the rejection of positivism: it is sometimes said that there simply *are* no such things as reasons or evidence in any important sense. For example, even my story of how rules of reasoning and method arise, and of how, despite the rejection of positivism, there is a "given" in observation, might be rejected as not providing the required objectivity and authority to count as *really* being reasons or evidence. Such arguments, I think, are part of the reason for the widespread neglect, by historians of science especially, of "final-form" language of science as somehow irrelevant to an understanding of what goes on in science.

I also think those contentions can be shown to be incorrect, but that is another story. Here my main aim has been to show that it is possible to understand what scientific language itself is, without assuming positivistic universal and necessary rules, and why that language is itself supposed to be persuasive in a normative sense. A full account of the extent and limits of the persuasive power of scientific language would require a discussion of the contention that such skeptical arguments cannot be convincing, not because of the character and emotional or ideological attachments of human beings but because they are intrinsically inadequate to establish their claims. That discussion cannot be given here; but in conclusion, I would like to offer a brief comment (a piece of rhetoric, perhaps, but in the sense I have called scientific) designed to persuade you that the problem is in dire need of reconsideration: that it may be a profound and even (for philosophers, historians, and sociologists of science at any rate) tragic mistake to be supposing so widely and confidently that scientific reasons and evidence are undetermined and afflicted by the necessary existence of Duhemian alternatives.

I mentioned earlier that there is by now very strong reason to require that theories in those fields must be constructed within the framework of the combined Standard Model-Big Bang theory. Though incomplete in specifiable respects, the Standard Model is highly successful in accounting for the nongravitational features of the world around us today: there exists no experimental result that contradicts it. When applied to Big Bang cosmology, it provides a highly detailed picture of what occurred in the universe from about 10^{-35} seconds after the Big Bang: for example, of the breaking of the unity of the strong and electroweak forces, and later of the electromagnetic and weak forces; and of the production of primordial helium and deuterium nuclei. The predictions that result are also highly successful. Big Bang cosmology is not as secure, but it remains an excellent theory, especially in its Standard-Model formulation. That formulation offers the framework for a unified understanding of the history of the universe, and to an extent depending on whether a deeper theory than the Standard Model and its cosmological application can be arrived at, it may ultimately give such a picture from the beginning of the universe to its fate in the far future. The picture is a coherent one: masses of gas in the expanding universe collapse to form galaxies and their components; gravity creates conditions in which stars cook primordial hydrogen and helium into higher elements by nuclear reactions, spewing them out into space, sometimes exploding and producing still heavier elements, all becoming available in interstellar clouds for future generations of stars; those future generations can produce planetary systems, in which, under certain conditions, self-reproducing molecules may arise and life evolve. There are gaps (though not of the Duhemian sort) at many stages—how the density fluctuations that become galaxies arise, the exact circumstances under which stars with planetary systems form, how complex self-replicating molecules could have arisen. But in none of these cases, except possibly the first, is there any present reason to believe that the gap is anything more than just that, a gap in our present understanding of a total unitary picture, a gap that like that concerning the birth of stars can in principle be filled. Furthermore, the constraints on further development of Standard-Model cosmologies—in the directions of higher group-theoretic models, supersymmetry, and string theory, for example—have grown tighter and tighter. Gauge theories are renormalizable; in general, field theories are not. Only with local gauge theories do the known forces of nature arise naturally and inevitably. Only with non-Abelian local gauge theories, do we obtain characteristics required to account for the nature of the strong force. Only with supersymmetry do we seem to have hope of unifying quantum field theories with gravity. Only in string theories do we get stringent limits on the group structures that might do justice to the universe we know, and only in such theories do we escape mathematical difficulties (anomalies in a technical sense, and, hopefully, various sorts of infinities) that have plagued previous theories. Though large numbers of alternative possibilities seem to sneak in through the back door even with such constraints, those alternatives are, again, not of the sort envisioned by Duhem and Quine, and many scientists express hope or even confidence that new constraints will yet be found, and even that an ultimate theory will be so tightly constrained that no alternative is possible. What I want to point out is the sad irony involved: that at the very time when science is approaching the attainment of a unified, unique theory in a way it has never approached before, philosophers, historians, and sociologists should be questioning whether such uniqueness and unification can in principle be achieved, and whether alternatives can ever be ruled out. It seems that philosophers and others are missing what's going on, and, as they have done so often before to their regret, even denying the possibility that it could happen. This is a major consideration that I hope will persuade at least some to rethink the grounds for skepticism about the potentialities of science. Even though that rethinking cannot return us to the absolutes of positivism, it must take into account the ways in which science, and its expression in language, has been built on a long history of conclusions at which it has previously arrived.

But as I have said, that rethinking must be a topic for another occasion, and with it also a full assessment of the conceptions of the role of rhetoric that accompanies the skeptical arguments to which I have referred. Here my purpose has been to show that it is possible to give an account of what scientific language is, and how it develops, without assuming universal and/or necessary rules, and why statements expressed in that language can properly be held to be persuasive in a normative sense. In speaking of "the language of science," I remind you again that I am speaking of final-form language, the language in which, for the reasons I have discussed, it is considered most appropriate to express scientific work. In the past two decades or so, such finalized, polished statements have been widely ignored by philosophers, historians, and sociologists of science. It is said, for example, that the "real work" of science comes in what leads up to the finalized statement; that the finalized form masks the processes that go into arriving at that form; and that once one understands the prior work, the form of the finalized statement falls out more or less automatically. All this is true, and even important in some respects for some purposes; but as I have argued, some prior work is more relevant scientifically, in a clearly specifiable sense, and those aspects of the prior work are embodied in, presupposed in, the finalized statement, as constituting the scientific background of that formulation. This was what we saw in the case of the background of the Lada-Shu formulation of the objective of modern star-formation theory. Problems, lines of research, and possibilities expressed in such language achieve a certain independence of their human origins and their human users. For regardless of the convoluted paths by which they are arrived at, once their final form has been arrived at and accepted, they are at hand to serve as the primary scientific background functioning in the further development of science. The neglect of those factors, the failure to distinguish them from others (social, institutional, personal, rhetorical) are therefore clearly unfortunate. For whatever ultimate status is accorded to what I have described as the evolved, internal, contingent conclusions of science, whatever the ultimate validity of the rules that those conclusions engender, the final-form expression of scientific ideas and its scientific background, and the role of that background as a factor in shaping the course of science, must at least be taken into account in asking about the role of rhetoric in the scientific enterprise. Its understanding is, indeed, as I have tried to show, a prerequisite for trying to understand the role of rhetoric. This is especially true when we turn our attention to what is actually occurring in the science of our present age.

NOTES

1. C. J. Lada and F. H. Shu, "The Formation of Sunlike Stars," *Science*, 248 (4 May 1990): p. 564.

2. D. Shapere, "The Concept of Observation in Science and Philosophy," *Philosophy of Science*, 49 (December, 1982): pp. 485–525.

3. E. A. Havelock, "The Linguistic Task of the Presocratics," in *Language and Thought in Early Greek Philosophy*, K. Robb, ed. (La Salle, Ill., Monist Library of Philosophy, 1983), pp. 7–82.

4. Aristotle, *Rhetoric*, in *The Basic Works of Aristotle*, R. McKeon, ed. (New York: Random House, 1966), p. 1330 (1356a25).

5. *Ibid.*, (1356a26).

6. Letter from Marcello Pera, Sept. 22, 1989.

BIBLIOGRAPHY

Aristotle. *Rhetoric*, in *The Basic Works of Aristotle*, ed. by R. McKeon (New York, Random House, 1966), pp. 1325–1451.

Havelock, E. A. "The Linguistic Task of the Presocratics," in *Language and Thought in Early Greek Philosophy*, K. Robb, ed. (La Salle, Ill.: Monist Library of Philosophy), 1983.

Lada, C. J. and F. H. Shu. "The Formation of Sunlike Stars," *Science*, 248:564–572 (4 May 1990).

Shapere, D. "The Concept of Observation in Science and Philosophy," *Philosophy of Science*, 49:485–525 (December, 1982).

PART II

Rhetoric in Action

Galileo and Newton: Different Rhetorical Strategies

RICHARD S. WESTFALL

I need to open, not with an apology, but with an explanation. After a brief hesitation, occasioned by the fact that I have never formally studied the discipline of rhetoric, I accepted the invitation to participate in a conference on "Rhetoric in Science." It follows from my hesitation that when I speak of "rhetoric" I employ a wholly intuitive, common sense understanding of the word. What I bring to compensate the deficiencies in my knowledge of rhetoric is extended study of Galileo and Newton, and more recent but nevertheless intense investigation of the scientific community of the sixteenth and seventeenth centuries. To these I join the conviction, an extension of my common sense understanding of rhetoric, that every author, including the author of a scientific work, is addressing some audience that he believes to exist.

Let me pause briefly to examine that last statement. As the author of several books, I cannot remember having ever systematically attempted to define the audience to which I was writing, and I tend to suspect that neither Galileo nor Newton consciously tailored their books to specific groups. On the other hand, I have always written in the knowledge that there is an extensive community of historians of science, as well as cultural historians, scientists, and others, whom I can expect to be interested in studies of science in the seventeenth century. I assume then that Galileo and Newton, in the same way, without consciously defining an audience, wrote in the knowledge that there was someone out there to address, and that they adopted rhetorical strategies suited to their respective scientific communities.

I turn to Galileo at the beginning of the seventeenth century, well over 50 years after Copernicus' book but still in the earliest phase of the Scientific Revolution. It was Galileo's enormous good fortune to be able to introduce himself to the larger public, beyond the limited circles in which his wit and brilliance had made his presence known, with a book that could not fail to command attention, the *Starry Messenger*, his announcement that here are bodies in the heavens never before seen by any human being. The *Starry Messenger* sustained Galileo's reputation until the end of his life and insured that nothing he published would be ignored. From the beginning, however, Galileo had more in mind than conveying messages from new celestial bodies. The *Starry Messenger* explicitly attached the new bodies to the Copernican system.[1] The *Letters on Sunspots* three years later were more outspokenly Copernican.[2] The intervening *Discourse on Bodies in Water* had made it clear meanwhile that the goal Galileo pursued transcended the Copernican system; it was nothing less than supplanting established Aristotelian natural philosophy.[3] If the announcement of new bodies in the heavens was sufficiently sensational as not to require a rhetorical strategy, the creation of a new natural philosophy demanded both a community prepared to participate in the enterprise and a rhetorical strategy to communicate with them.

The strategy appears to me to be embodied in the very nature of Galileo's works. Although he is rightly received as the fountainhead of modern mathematical science, Galileo's books were not forbiddingly technical. As every historian of science knows, undergraduates today can read and appreciate Galileo as they cannot read and appreciate, for example, Newton—or Kepler, or Huygens. The *Discourses* established the modern science of kinematics and became the foundation on which scientists of the seventeenth century constructed a new physics. For all of that, the *Discourses* were not oppressive in their technical demands. The mathematics involved was the simple geometry of triangles, rising at its greatest level of complexity to the definition of the parabola. The seventeenth century was an age of enormous mathematical activity, and there was no shortage of persons able to cope with geometry at this level. As I said before, I see no need to insist on a conscious choice of strategy in this matter. Galileo himself was not a great mathematician, and the level of mathematics in his works reflected his own capacity. Moreover, the geometry he employed was adequate to his purpose. His originality lay not in the level of the mathematics but in the new program to which he applied it. The need was to introduce people to a different approach to the science of motion. Figures of speech like those Butterfield introduced seem necessary here. Galileo was trying to get people to put on a new thinking cap, to pick up the stick at the other end, to see things from a new perspective.

From this point of view, we can appreciate Galileo's repeated reference to ships, especially in the *Dialogue*, his most important polemic, not just for the Copernican system, but for the broader new program. I refer not so much to the well-known discussion of how a stone falls from the mast of a moving ship, the issue to which Koyré gave prominence, but to shorter passages that would have spoken more directly to remembered experience, one concerned with goods on a vessel that travels from Venice to Aleppo, another with butterflies, incense, dripping water, and games of catch in the closed cabin of a ship.[4] Galileo was addressing most immediately the great sea-faring people of medieval and early modern Europe. To them he said in effect, "You have been on ships. Stop and reflect on what you did in fact see even though you may not have realized you were doing so. There was not 'the least change in all the effects named, nor could you tell from any of them whether the ship was moving or standing still.' A science of motion that does not embrace events on a moving ship as well as those on dry land cannot be adequate. We must learn to see phenomena of motion in a new way." A similar analysis, which looked toward the reordering of another range of common experience, could be offered of Galileo's discussion of light in Day One of the *Dialogue*.

To whom was this message directed? In my understanding of the world, a call for a radical reordering of experience, with its overt implication that the existing system of natural philosophy is mistaken, is most unlikely to be received happily by an established scientific community. There was such a scientific community in Galileo's day, and we should be mistaken to refer to it by any other name. It stretched back to the foundation of the medieval universities and the recovery of Aristotle in the late twelfth century. Galileo spent the first 21 years of his professional life as part of that community. As a young professor at Pisa he devoted considerable energy to compiling commentaries on Aristotle's *De caelo* and his logic from the lecture notes of Jesuit philosophers at the Collegio Romano.[5] In Padua he made the acquaintance of Cremonini. Soon after Galileo returned to Florence, he became embroiled with the resident Scholastic philosophers on the issue of why ice floats; four peripatetics, lead by Ludovico delle Colombe, entered the lists against him.[6] Galileo knew the established community of Aristotelian natural philosophers well. Nothing in his work suggests that he expected to convert them to his new perspective.

Indeed, he devoted a good part of his rhetorical strategy to casting scorn upon them. Satire directed at the Scholastics is one of the common threads that runs through the entire fabric of Galileo's output. Already his early *De motu*, which grew directly out of the commentary on *De caelo*, took the form of a tirade against Aristotle. "Heavens!" he exclaimed, toward the end of a discussion of heavy and light. "At this point I am weary and ashamed of having to use so many words to refute such childish arguments and such inept attempts at subtleties as those which Aristotle crams into the whole of Book 4 of *De Caelo*. . . ."[7] The current of disdain continued through Galileo's published works, sometimes more muted than in other places, but never absent. Referring less to Aristotle than to latter-day peripatetics, it supplied the fundamental rhythm to the *Assayer*, rising at points to exclamations of open contempt.

> It seems to me that I discern in Sarsi a firm belief that in philosophizing it is essential to support oneself upon the opinion of some celebrated author, as if when our minds are not wedded to the reasoning of some other person they ought to remain completely barren and sterile. . . . Sarsi perhaps believes that all the hosts of good philosophers may be enclosed within walls of some sort. I believe, Sarsi, that they fly, and that they fly alone like eagles, and not like starlings. It is true that because eagles are scarce they are little seen and less heard, while birds that fly in flocks fill the sky with shrieks and cries wherever they settle, and befoul the earth beneath them.[8]

"Oh, the inexpressible baseness of abject minds," Galileo was equally shrill in the *Dialogue*.

> To make themselves slaves willingly; to accept decrees as inviolable; to place themselves under obligation and to call themselves persuaded and convinced by arguments that are so 'powerful' and 'clearly conclusive' that they themselves cannot tell the purpose for which they were written, or what conclusion they serve to prove![9]

Again, I am not convinced that this was a consciously formulated strategy. Galileo had a natural gift for sarcasm and satire, and it seems clear that he delighted to use it. The *Discourse on Comets* is especially revealing. On a question of astronomy raised in the very shadow of the decree of 1616, Galileo was unable to rein in his penchant for savage wit. He went out of his way to begin with a gratuitous insult addressed to Scheiner, cast in a form that recalls his thrusts at peripatetics in general. Scheiner pretends to be Apelles, though in fact he "could not compare in skill even with the most mediocre painters."[10] Although Galileo did not direct a barb this pointed at Grassi, he made his disdain sufficiently obvious.[11] Thus he succeeded in converting the formidable Jesuit order, his erstwhile friends and supporters, into implacable enemies. For all that, Galileo did intend that his books should convince, and it is highly unlikely that he would have adopted such language had he not believed that there was an audience prepared to respond.

We know from other sources that Anti-Aristotlelianism did have an audience. The spread of Neoplatonic and hermetic philosophies during the sixteenth century testifies as much. In studying the scientific community during that age, I have run into a number of individuals to whom Scholastic philosophy had ceased to carry conviction, to whom apparently it had begun to sound like the lifeless repetition of phrases repeated too often. Galileo's friend Ciampoli was one of them, though Ciampoli may have learned to nurse these sentiments when he met Galileo at the Florentine court.[12] Descartes's description of his education in Part I of the *Discourse on Method* offers the classic statement of this attitude.

It appears likely to me that similar sentiments were more common within aristocratic circles. If such men went to a university at all, it was for polish and entertainment, and they seldom bothered to complete a degree. They had no intention of making a career out of philosophy. They would have been prone to look down on academic philosophers who came for the most part from a lower order of society. Members of the ruling class at various levels acted as patrons to the radical philosophers of the sixteenth century, and when a vociferous opponent of Aristotle, such as Ramus, was driven from the University of Paris, he found refuge with high aristocrats.

Galileo knew the aristocratic patrons. Though impoverished, he was of their class. He populated his dialogues with their representatives, not only Salviati and Sagredo, but in the so-called Sixth Day of the *Discourses* Paolo Aproino and Daniello Antonini.[13] Beyond the personae of the dialogues, there were others, such as Federico Cesi and Virginio Cesarini in Rome and in Florence Mario Guiducci, Antonio Nardi, Pier Francesco Rinuccini, and the two Arrighetti cousins, Niccolo and Andrea.[14] Galileo's use of the vernacular, in contrast to the Latin of university discourse, may well have been related to the anti-Aristotelian strategy. Although Italian aristocrats knew Latin perfectly well, they had abandoned it in favor of the vernacular in administering the states of Italy. It seems likely to me in any case that Galileo utilized his satire of Scholastic philosophy as one device to gain a hearing among this class.

Nevertheless, I cannot believe that this was the whole of Galileo's strategy. If his goal was to supplant Aristotelian natural philosophy, he must have known that he could not effect that end with aristocratic dilettantes, whose real concerns did not lie with philosophy. Filippo Salviati and Giovan Francesco Sagredo were undoubtedly intelligent men, but they live today through Galileo's dialogues and not through any contributions of their own to science or learning. As astute an observer as Galileo must surely have understood their limitations.

Who then was the principal audience he addressed? To understand Galileo's strategy in relation to the scientific community in the early phase of the scientific revolution, I think we have to realize that at the time Galileo returned to Florence, that is, at the time he began to publish, the audience was virtually nonexistent. A passage in Isaac Beeckman's diary at much this same time, when Descartes reappeared in the Netherlands in 1628 and came to renew their friendship, is relevant. In 1618 the two men had found that they shared an outlook in natural philosophy, an outlook similar to Galileo's. When Descartes returned to the Netherlands, he told Beeckman that in ten years of travel through Europe

> he had not found anyone else with whom he could discuss his concep-
> tion of things [*secundum animi sui sententiam*] and from whom he
> could hope for assistance in his studies. Everywhere he found a dearth
> of true philosophy, which he calls the work of the competent [*operam
> navantium*].[15]

That is, the creation of a new natural philosophy demanded the creation of a new scientific community outside the ranks of the Aristotelian philosophers in the universities, whom Galileo never expected to convert. Galileo's rhetorical strategy was directed primarily to this task.

However—and this has appeared ever more critical to me as I have begun to think about rhetorical strategy and the community to which it was addressed—Galileo had good reason to think that the creation of such a community was possible, that young men not yet coopted into the entrenched community of Scholastic philosophers could be enlisted in his program. For already he had enlisted at least one such follower in Padua, Benedetto Castelli. Let me repeat once more that I do not find it necessary to contend that Galileo consciously formulated a strategy like the one I have stated. It does seem essential to me that a person like Castelli existed, living proof that someone would listen to what he said. Castelli was not alone; among the other students we can name were Antonini and Aproino. They were more in the mold of Sagredo and Salviati than of Castelli, however, and though their correspondence with Galileo in the period after he returned to Florence indicated their conversion, they did not stay the course as Castelli did. [16]

At the first appearance of Castelli in Galileo's correspondence, only a few months after Galileo's return to Florence, he made it clear that he considered himself Galileo's follower. Perhaps "client" is the better word. Castelli was desperate to escape from the monastery in Brescia into the world of science to which Galileo had introduced him, and he appealed to Galileo for assistance. [17] By the fall of 1613, Galileo had installed him as a professor of mathematics in the University of Pisa. For his part Castelli was explicit in stating his conversion to Galileo's program. "You who raised me up out of abject ignorance," he later told Galileo, in reference to a further possible appointment, "can affirm with full assurance [*a buona chiera*] that I know my trade and that my appointment will be wise. If I speak too freely, excuse me that with you, who is to me father, master, and patron, I should behave in this manner." [18]

Castelli continued as a constant factor in the rest of Galileo's life. The master exploited him egregiously. He made Castelli collect his stipend, which was paid in

Pisa as part of the university's budget, and even utilized him to importune the authorities for advances.[19] When Galileo's son Vincenzio came to study in Pisa, Castelli had to function *in loco parentis*, and later in Rome, when Galileo's nephew, another Vincenzio Galilei, arrived there to study music, Castelli assumed the same role for him. As it turned out, neither young man was interested in learning, and in shepherding them through a series of disasters, Castelli was forced to pay a stiff price for his discipleship. "Tell me now what sin I have committed," he finally exploded in exasperation, "that these two scapegraces have put me through purgatory."[20] Nevertheless, even though the summons to Rome under the patronage of the Barberini had rendered Castelli no longer dependent on Galileo materially, he never hesitated in offering his service. He devoted endless time to the details of the pensions that Urban bestowed on Galileo and his family. Even after the trial, when concern for Galileo cannot have endeared him to the papal court, Castelli worked for his liberation.

And what matters most, Castelli functioned always as an active part of Galileo's program of investigation and thus as a constant reminder that an audience, whose opinion he valued, was prepared to receive his message. Castelli observed the satellites of Jupiter for the master, transmitting to him a body of data to help in defining their periods.[21] Later on he tested Galileo's binocular device, designed to make it possible to observe the satellites at sea and thus to determine longitude; he paid for his devotion this time with seasickness.[22] Castelli suggested to Galileo the observation of the phases of Venus and the means to make detailed observations of sunspots by projecting the sun's image through a telescope onto a screen.[23] He composed most of the *Risposta* to Galileo's opponents on bodies floating in water, but put himself forward in the publication only to the extent of signing the dedication.[24] At some personal risk he stood in Galileo's stead to defend astronomy from theological assault before the Grand Duchess Christina, and later he probably helped furnish Galileo with the passages from the fathers of the church that he used to buttress his argument in the famous *Letter to the Grand Duchess*. After the decree of 1616, he even made observations clearly intended to reveal stellar parallax and thus presumably to demonstrate that nevertheless the earth does move.[25]

Through it all Castelli constantly made it clear that he was Galileo's disciple. When Galileo insisted on paying the expenses necessary to clear the pensions granted in Rome, Castelli grudgingly agreed. "As far as the debt that you owe me is concerned, I am ashamed to reply, because I am the debtor, and I can never pay it."[26] If not before, Galileo could not have failed to be convinced of his follower's complete conversion once he began to find in him his own spirit of mocking sarcasm. Early on, Castelli told Galileo of a priest from Genoa who had been convinced of the truth of Copernicanism by its opponents. One of them had told him that if Copernicus were correct, he would see the door to his room every morning in a different position from where it had been the night before. As the priest had concluded, "it is impossible that the earth be at rest if men such as that believe it is."[27] More than 20 years later, Castelli was still treating their opponents in the same manner. A young scholar in Rome had come to him to report that when a brick was exposed to the sun with half of it painted black and the other half white, the black half became hotter. The scholar was also a student of an eminent peripatetic philosopher at the Gregorian College. Castelli instructed him to take the case to the philosopher but to tell it to him in reversed form, that is, to tell the philosopher that the white end became more hot. The philosopher, Castelli assured the scholar, would explain why this must happen, and so in fact he did. After Castelli and the scholar had had a good laugh, they tried the experiment again, in good Galilean style, to be sure of the result; the black half did indeed get much hotter. The scholar then returned to the philosopher, who after thinking at first that the student had confused the results, finally called upon his highest and most recondite speculations to explain this as well. Castelli stood in awe, he assured Galileo, "of a mode of philosophy so subtle."[28] The master could not have dismissed it more caustically himself.

Meanwhile, Castelli the convert had also become the apostle of the new truth, and through him Galileo acquired another significant follower, Bonaventura Cavalieri. Cavalieri had come to Pisa to study mathematics with Castelli. Through Castelli he had established a connection with Galileo, and almost immediately he had begun to rely, as Castelli had done before, on Galileo's patronage to pursue a career in mathematics. After Cavalieri had been called home to Milan in 1620 to teach theology in a monastery of his order, the Jesuits, he maintained what appears to have been a one-sided correspondence with Galileo, who was his link to the studies he loved. His letter of 28 July 1621 was strikingly similar to the one from Castelli a decade earlier. Cavalieri found himself isolated among people who did not understand his studies and kept asking him what profit there was in mathematics; desperately seeking Galileo's assistance to escape, he announced himself as the disciple of Galileo's philosophy, "which deserves to be set before all others, in so far as it offers a very natural portrait of nature, whereas others are similar to images reflected in water that is greatly ruffled. . . . "[29] Eventually, after Cavalieri had suffered more than one disappointment, Galileo was instrumental in his appointment to a chair in mathematics at the University of Bologna. [30]

Cavalieri consciously timed his publications to coincide with the years in which he would need to be reappointed at Bologna, and it was his misfortune that the timing coincided with the opening stages of Galileo's trial. *Lo specchio ustorio*, dedicated to the Senate of Bologna in 1632 in gratitude for his reappointment, contained a demonstration that the trajectory of a projectile is a parabola. Galileo received the information about Cavalieri's book soon after the Church had suspended the distribution of his *Dialogue*, and burdened already with too many troubles, he flew off the handle and accused Cavalieri of stealing his discovery. Cavalieri could not have been more distressed. He had thought, he assured Galileo, that he was acting out the role of a faithful disciple, spreading the word of the master's discovery, which he had specifically identified in the book as Galileo's work. He offered to withdraw *Lo specchio ustorio* from circulation and to delete the offending passage, so that Galileo would not have to say, with Caesar, "Tu quoque, Brute, fili!" "I have always considered it my greatest fortune to have known you and to be able to honor and serve you, as I enjoy the precious fruits of your eminent learning. . . . "[31] The tempest quickly blew itself out as Galileo regained his composure; his warmest recommendations of Cavalieri came after the brief storm. No more than Castelli did Cavalieri allow the trial to interrupt a relation that he found central to his existence. "I know," he assured Galileo, "that in relation to you I am a shadow, which therefore continues to follow, at least with desire, the movement of the brilliant light that shines in you creating the shadow."[32] It is worth recalling, as we consider Galileo's rhetorical strategy, that it was not the *Dialogue* and the *Discourses* that effected Cavalieri's conversion. Like Castelli, he enlisted as Galileo's disciple even before the *Assayer*.

Other disciples were joining them in a nascent community well before the publication of the *Dialogue*. In 1626 Castelli was called to Rome, leaving the chair in Pisa vacant. Cavalieri hoped to win the appointment, but it went instead to Niccolo Aggiunti, a tutor to Ferdinand II who had also studied mathematics with Castelli in Pisa. Although he did not gain the chair as Galileo's client, Aggiunti did enroll as his disciple, and he later hoped that Galileo's support would lead to an appointment in Padua.[33] He sought Galileo's opinion about the design of a bridge.[34] In 1630 when the completed *Dialogue* was being read in manuscript and debated in the Tuscan court, Aggiunti made his voice heard in Galileo's support,[35] and later on in the perilous days of February 1633, he was sufficiently devoted that he went to Arcetri with Geri Bocchineri, an official in the Tuscan court who was the brother-in-law of Galileo's son, to remove Galileo's papers to safety.[36]

When Aggiunti died in 1636, Dino Peri, who also had a powerful patron, Cardinal Capponi, received the chair of mathematics in Pisa. A friend of Aggiunti, he had already committed himself to Galileo's camp. In 1633 he and Aggiunti had been engaged in complex negotiations that called upon Galileo's support, whereby Aggiunti would move to Padua and Peri would assume the position in Pisa, and at that time, on the very morrow of the trial, Peri had announced their intention to see the school of Galileo established, despite its persecutors, in all the universities of Italy.[37] During 1637 Peri spent at least one period at Arcetri to help as Galileo's eyes began seriously to fail.[38] That same year another young man, Vincenzo

Renieri, who was skilled in observation, appeared on Galileo's horizon. He entered seriously into the program of defining the periods of the satellites of Jupiter and perfecting the method of determining longitude, and he spoke of established philosophy, in the accents of the master, as "that cursed servility to Aristotle."[39] When Peri died, Renieri succeeded him at Pisa in 1640. Castelli, who had been invited but declined to return to Pisa at this time, suggested a student of his, Giovanni Alphonso Borelli, who would later figure in the heritage of Galileo in Tuscany and who should be considered a member of the growing community in 1640.[40] Galileo also assisted the appointment of Alessandro Marsili to a chair in philosophy at Pisa.[41]

Before this time another student of Castelli, Evangelista Torricelli, had announced his adherence to Galileo's point of view. Writing from Rome in September 1632, Torricelli indicated that he had read the *Dialogue*, which he praised even though the Church had already suspended its circulation. "I consider myself most fortunate in this, that I was born in a century when I could know a Galileo, that is, an oracle of nature, and revere him by letter. . . ."[42] Nearly ten years later, Torricelli sent Galileo a treatise on motion founded upon Galileo's own work,[43] and as everyone knows, he was able to revere Galileo briefly in the flesh, joining him in Arcetri a few months before the master's death.

In Arcetri Torricelli found Vincenzo Viviani, a young man of promise whom the court had sent to Galileo in the fall of 1639 to assist him in his blindness and to profit from his instruction.[44] Viviani had insisted on the revision of a crucial proposition in the *Discourses* and had entered actively into the process of doing it. In the years ahead, after Galileo's death, he would become his quintessential disciple.

Before Viviani appeared on the scene, Galileo had made another convert, Fulgenzio Micanzio, the follower of Sarpi in Venice, not really a natural philosopher but nevertheless a man who became increasingly excited about Galileo's work in the years following the trial. When he wrote in 1634, Micanzio, a cleric, had been studying the *Dialogue*, quite unmindful of the ban the Church had placed on it. It seemed to him "that when the peripatetic principles are reexamined, as you have done in regard to the constitution of the universe, they all go up in smoke before me."[45] Micanzio began to explore the possibility of republishing all of Galileo's works, though he quickly discovered that the Inquisition would not allow it. He did receive the manuscript of the *Discourses*, chapter by chapter as it were, as Galileo completed the book, and he entered into the tortuous plans to publish it abroad. Things of such value must not be allowed to perish, he told Galileo,

> and they are such that, before God and my conscience, I believe them
> to be the greatest step forward in philosophy that has been taken

during the past two thousand years, and defrauding the world of it would be a crime against humanity.[46]

In 1634 another group of young followers also entered Galileo's life, the Padri delle Scuole Pie, or Fathers of the Pious Schools. One of them, Fabiano Michelini, who was tutor to Prince Leopold in mathematics, began to function as a courier between Galileo and Castelli, who feared that letters through normal channels would be intercepted. In 1638 he forwarded the prince's request for a demonstration of one of Galileo's propositions on motion, and a few months later, he sent Galileo a demonstration of his own, which "I know to be wholly from you," for comment.[47] In addition to Michelini, there were Fathers Clemente Settimi, who helped Galileo in his blindness and obtained permission to spend nights in that service at Arcetri away from his monastery, and Ambrogio Ambrogi. At least Settimi was in correspondence with Renieri, and together with another young Galilean, Marsili, the Fathers of the Pious Schools formed a circle around Prince Leopold that appears to have been a virtual *Accademia del Cimento* in embryo. In March 1640 Leopold wrote to Galileo on behalf of the circle about Liceti's book on the Bolognian stone and his related objection to Galileo's opinion about the secondary light of the moon. In response Galileo composed an extended essay on the subject that he addressed to the prince.[48] The group also studied Galileo's *Discourse on Bodies in Water*, and later that year, Galileo sent works of Archimedes to Leopold, presumably at the request of the circle.[49] The circle was explicitly enough Galilean to be denounced to the Holy Office.[50] As I suggested above, it was the prototype of the later *Accademia del Cimento*.

It has been customary to treat Federico Cesi's *Accademia dei Lincei*, of which Galileo was a member, as the first modern scientific society. I think that this is a mistake and that the *Accademia* is better understood as the vehicle that Cesi created for his patronage of learning. The group of followers of Galileo that had come into being by 1640, a group without formal organization though with multiple mutual connections, seems to me much more like a scientific society, albeit still a nascent one, or perhaps better a nascent scientific community. It was not alone. Only a few years later, a similar one came into being, initially in the Netherlands, in response to Descartes' publications. I do not mean to suggest that Galileo and Descartes alone led the rebellion against Aristotelian philosophy, or that they evoked responses only in Italy and the Netherlands. By the middle of the century, from a number of sources, the modern scientific community was beginning to take shape. A significant segment of it arose in response to Galileo's rhetorical strategy, that is, in response to his success in expressing what he had to say in a form that a number of young natural philosophers were prepared to adopt as their own.

When we compare Galileo to Newton, some of the factors appear to be the same. Newton also led a rebellion against an established natural philosophy, and Newton also gathered around him a circle of disciples, for the most part young disciples, as Galileo's had been, who were committed to the new order. Nevertheless, when we examine the situation closely, the apparent similarities dissolve, and the issues of rhetorical strategy and the communities to which they were directed present instead contrasts between the two men.

I start with the rhetoric itself. What could be more different? Galileo's writings are read today in Italian schools as masterpieces of the national tongue. Even if Newton had chosen to write his major work in English, one cannot imagine its being read in a similar way today in schools of the English-speaking world. Although Newton commanded an effective style that clearly conveyed his meaning, the *Principia* contains more mathematics than prose. The mathematics, moreover, could not be more forbidding. The difficulty is in fact even greater than it appears on the surface. The *Principia* seems to proceed in terms of geometric proportions, albeit extremely complex ones. When one seriously wrestles with the demonstrations, however, one soon recognizes that one is not dealing with classical geometry. The method of nascent and ultimate ratios, that is, the instantaneous variations of quantities at points on a curve, merely employs the idiom of geometry.[51] The thought patterns that are being expressed are those of the calculus. Few indeed are the books that have compromised less with readability.

Newton's other great book, the *Opticks*, does express itself in prose rather than mathematics. However, the *Opticks* is not easy going. Newton presented a carefully conceived program of experimentation, which did not call upon common experience at all, but demanded such care that French scientists, for example, did not accept the basic experiments for nearly half a century after their initial publication in the article of 1672 because they could not reproduce them.[52] Clearly we are dealing with a different rhetorical strategy.

There is another striking difference. We do not find in Newton's published works anything that corresponds to the stream of invective that Galileo poured over the Scholastics. It was not that Newton was tolerant of Cartesian natural philosophy. Quite the contrary, more than merely rejecting Cartesianism, Newton felt for it a deep antipathy. If one learns to know this from his manuscripts, one can catch glimpses of it here and there in the published works,[53] but it does not break out in the uninterrupted carnival of disdain that characterizes Galileo's writings. That is, Newton sought to address the current school of natural philosophy instead of dismissing it. As with Galileo, I do not mean to imply that Newton consciously chose a strategy. I am prepared to state that the rhetorical strategy inherent in his work aimed to convert an existing scientific community instead of driving it into the wilderness.

The problem Newton confronted differed fundamentally from that of Galileo.

As far as the world of science was concerned, the generation before him had effectively dismantled the qualitative philosophy of Aristotle that ordered daily experience. Newton met the peripatetic philosophy when he first arrived in Cambridge. Sometime during his undergraduate years, he discovered the literature of the new science—Galileo, Descartes, Gassendi, Boyle, and others. [54] Later, and importantly, there would be Huygens. From the time of his first contact with them, these writers defined natural philosophy for Newton. The testimony of his notebook seems clearly to indicate that he made the transition almost instantly and with ease. It left behind no memory of a bitter struggle that demanded expression. As far as Newton's science was concerned, Aristotelian natural philosophy might never have existed. He never referred, even in derision, to the reading of his first years in Cambridge. The problems with which he dealt were problems within the new tradition.

Unlike Galileo then, he did not face the need to create a new scientific community. Rather he could address the community that Galileo and others of his generation had created, a community Newton knew to exist. And he could address them in the tongues he knew they used, mathematics and experimentation. There is every reason to believe that most scientists at the end of the seventeenth century found the mathematics of the *Principia* extremely difficult, and I have pointed out how delicate the optical experiments were. For all that, mathematics and experiment had been installed as the idioms of the new science, and it appears that the members of the scientific community received Newton, not as an enemy who assaulted their standards, but as one of their number who raised those standards to a higher level of perfection, which became henceforth the model they would struggle to emulate. Certainly this was Newton's reception in Britain and not much less in the Netherlands. It is true that his concept of attraction at a distance raised obstacles, especially in France, but only a decade after publication of the *Principia*, when France reorganized its *Académie des Sciences*, the academicians chose Newton as one of the eight foreign members. [55] The correspondence of the early 90's between Leibniz and Huygens, two scientists for whom the concept of attraction was anathema, reveals how impossible it was, even for those who found action at a distance inadmissable, to ignore a work that appealed so powerfully to other ideals they held as part of the new community. [56]

It follows also that Newton had no need for someone to play the role Castelli held in Galileo's life. Indeed, there does not seem to have been such a person. Halley came the closest, but again the differences outweigh the similarities. During the crucial years, when he discovered the calculus, wrote the *Principia*, and composed most of the *Opticks*, Newton was largely isolated. For him personal acquaintance did not help to define the audience he was addressing. He had met them rather in the books he read, which functioned for him as Castelli had for

Galileo, to convince him that an audience prepared to receive what he had to say
did exist. Galileo's *Dialogue* was one of the books. The works of Descartes were
others. The scientific community that Newton joined and addressed was the
community they and others of their generation had called into being.

When we turn, however, from the rhetorical strategies of Galileo and
Newton to the circles of disciples that gathered around each man, the similarities
here outweigh the differences. What the *Starry Messenger* did for Galileo early in
the century, the *Principia* did for Newton near its end. Both men, henceforth, were
figures to be reckoned with on the intellectual scene, and young men saw in their
influence means of advancement.[57] There is no need whatever to question the
sincerity of conversions, and there is no suggestion that desire for advancement
served only selfish and ignoble purposes. I mean merely to insist that a suitable
position enhanced the opportunity of a scientist to be heard, and that for a young
man without personal wealth, as was the case for the majority in each set of
disciples, a suitable position was the necessary condition for pursuing science at
all. It does appear to me that Newton's disciples, like Newton himself a generation
before them, were already members of the new community, and that they saw in
Newton, not an alternative to the now reigning tradition, but a brilliant enhance-
ment of it, a new definition as it were of excellence in the pursuit they had already
determined to follow. After reading the *Principia*, David Gregory wrote to thank
Newton "for having been at pains to teach the world that which I never expected
any man should have knowne."[58] In his review of the book for the *Philosophical
Transactions*, Edmond Halley insisted on its epochal significance:

> This incomparable Author . . . has in this Treatise . . . at once shewn
> what are the Principles of Natural Philosophy, and so far derived from
> them their consequences, that he seems to have exhausted his Argu-
> ment, and left little to be done by those that shall succeed him.[59]

This is the language of men speaking from inside the discipline they already
practice.

Seen from the perspective of rhetorical strategy, the Scientific Revolution
stands distinct from subsequent scientific revolutions. I trust that no one will be
inclined to suspect me of undervaluing the achievement of Isaac Newton. Monu-
mental as it was, it nevertheless differed in kind from the achievement of Galileo
and others of his generation. The emergence from the world of Aristotle into the
world of modern science was a unique event in Western history that set its stamp
on every subsequent development in the world of thought. Part of Galileo's genius
lay in his capacity to find a rhetorical strategy that convinced a number of able
young men to follow him across that divide. Perhaps Newton could have fore-

stalled Lagrange and remarked that Galileo had had the good fortune to cross the only divide that existed.

NOTES

1. *The Starry Messenger*, in *Discoveries and Opinions of Galileo*, trans. by Stillman Drake (Garden City, NY: Doubleday, 1957), pp. 34, 45, 57–58.

2. Perhaps it is superfluous to support this generally accepted statement, but see, for example, *History and Demonstrations Concerning Sunspots and their Phenomena*, in *ibid*. pp. 93–98, 112, 118, 142–144.

3. *Discourse on Bodies in Water*, trans. by Thomas Salisbury, Stillman Drake, ed. (Urbana, Ill.: University of Illinois Press, 1960). See especially pp. 19–26, where Galileo focuses the argument into a comparison of Archimedes, whose doctrine he finds true, "since it aptly agrees with true experiments," to Aristotle, "whose Deductions are fastened upon erroneous Conclusions."

4. The passage on the stone dropped from the mast: *Dialogue Concerning the Two Chief World Systems*, trans. by Stillman Drake (Berkeley: University of California Press, 1962), pp. 141–145; Venice to Aleppo: p. 116; phenomena in a closed cabin: pp. 186–188. Other references to appearances on a ship: pp. 61, 148, 155–156, 171–172, 249, 250, 251–253, 255, 375–376, 424–425. The Fourth Day, of course, brings the argument full circle and calls upon the motion of the earth to explain a phenomenon of the sea, the tides.

5. *Galileo's Early Notebooks: the Physical Questions*, trans. William A. Wallace (Notre Dame, Ind.: University of Notre Dame Press, 1977). *"Tractatio de praecognitionibus et praecognitis"* and *"Tractatio de demonstratione"*, trans. by William F. Edwards, William A. Wallace, ed. (Padova: Antenore, 1988).

6. *Bodies in Water*. The objections against him can be found in *Le Opere di Galileo Galilei*, Edizione nazionale, Antonio Favaro, ed. (Firenze: Barbèra, 1890–1909), vol. 4, pp. 145–244 and 313–440. His reply to them, *Risposta alle opposizioni*, is in *ibid.*, pp. 451–788.

7. *On Motion and On Mechanics*, trans. by I. E. Drabkin and Stillman Drake (Madison, Wi.: University of Wisconsin Press, 1960), p. 58. I have drawn the assertion that *De motu* grew directly out of the commentary on *De caelo* from the work of my student, Wallace Hooper.

8. A translation of the entire *Assayer* can be found in *The Controversy on the Comets of 1618*, trans. by Stillman Drake and C. D. O'Malley (Philadelphia: University of Pennsylvania Press, 1960). The present quotation is from pp. 183–189.

9. *Dialogue*, p. 112.

10. *Controversy on the Comets*, p. 24.

11. See, for example, *ibid.*, pp. 48–53.

12. See "Vita di Monsig. Giovanni Ciampoli Fiorentino," in Giovanni Targioni-Tozzetti, *Notizie degli aggrandimenti delle scienze fisiche accaduti in Toscana* (Firenze: Bouchard, 1780), vol. 2, p. 104.

13. *Opere*, vol. 8, pp. 319–346. Aproino enters the dialogue in the Sixth Day; Antonini is only mentioned as having participated in the experiments (p. 322).

14. For the relations of the two Arrighetti with Galileo and their knowledge of his mechanics, see their correspondence in December 1630 about the flood along the river Bisenzio, *ibid.*, vol. 14, pp. 176–202, passim.

15. *Journal tenu par Isaac Beeckman de 1604 à 1634*. C. de Waard, ed. (The Hague: Nijhoff, 1939–1953), vol. 3, p. 95.

16. See *Opere*, vol. 11, passim. Antonini, who pursued a military career, was killed in battle within a few years; one cannot assert with assurance what he might have done.

17. Castelli to Galileo, 24 Dec. 1610, *ibid.*, vol. 10, pp. 493–494.

18. Castelli to Galileo, 21 May 1617, *ibid.*, vol. 12, p. 318.

19. Castelli to Galileo, 6 and 13 May 1615, *ibid.*, vol. 12, pp. 177 and 179.

20. Castelli to Galileo, 24 June 1628, *ibid.*, vol. 13, pp. 437–438.

21. *Ibid.*, vol. 12, passim.

22. Annibale Guiducci to Galileo, 11 Sept. 1617, and Castelli to Galileo, 18 Sept. 1617, *ibid.*, vol. 12, pp. 344 and 346.

23. On the phases of Venus, see my article, "Science and Patronage. Galileo and the Telescope," *Isis*, 76 (1985), pp. 11–30. Galileo credited Castelli with the device to observe sunspots in his second letter about them, *Discoveries and Opinions*, p. 115.

24. *Opere*, vol. 4, pp. 451–788.

25. Castelli to Galileo, 16 Nov. 1616, 7 Jan. 1617, and 22 Feb. 1617, *ibid.*, vol. 12, pp. 296, 301, and 309. In 1627, I assume in connection with the composition of the *Dialogue*, Castelli again reported on such an observation, *ibid.*, vol. 13, p. 373.

26. Castelli to Galileo, 20 June 1631, *ibid.*, vol. 14, p. 277.

27. Castelli to Galileo, 12 Feb. 1614, *ibid.*, vol. 12, p. 26.

28. Castelli to Galileo, 27 June 1637, *ibid.*, vol. 17, pp. 121–123. Castelli was so delighted with the incident that he kept revising his exposition of it. Unfortunately, he did not possess Galileo's gift of style. Each version got longer and more pedantic, and the final essay had lost the incisiveness of the first telling. (Castelli to Galileo, 9 and 15 August 1937, *ibid.*, pp. 150–155 and 156–169.) He did assure Galileo that the original story was true. (Castelli to Galileo, 26 Sept. 1637, *ibid.*, p. 186.)

29. Cavalieri to Galileo, 28 July 1621, *ibid.*, vol. 13, pp. 70–72.

30. See Cavalieri's letters to Galileo from 24 Nov. 1628 to 20 Feb. 1629, *ibid.*, vols. 13 and 14. Also Galileo to Marsili, 12 Jan. 1630, *ibid.*, vol 14, pp. 65–66.

31. Cavalieri to Galileo, 21 Sept. 1632, *ibid.*, vol. 14, pp. 394–395.

32. Cavalieri to Galileo, 19 June 1635, *ibid.*, vol. 16, p. 281.

33. Aggiunti to Galileo, 17 April 1630, *ibid.*, vol. 14, pp. 94–95. See also Peri to Galileo, 24 Sept. 1633, *ibid.*, vol. 15, pp. 276–278.

34. Aggiunti to Galileo, 24 Jan. 1630, *ibid.*, vol. 14, pp. 69–70.

35. Aggiunti to Galileo, 24 Jan. 1630, *ibid.*, vol. 14, p. 70.

36. Aggiunti to Galileo, 27 Dec. 1633, *ibid.*, vol. 15, pp. 364–365.

37. Peri to Galileo, 24 Sept. 1633, *ibid.*, vol. 15, pp. 276–278.

38. Galileo to Guerrini, 9 May 1637, *ibid.*, vol. 17, pp. 75–76.

39. Renieri to Galileo, 27 Feb. 1637, *ibid.*, vol. 17, pp. 37–38.

40. Castelli to Galileo, 5 May 1640, *ibid.*, vol. 18, pp. 188–189.

41. Marsili to Galileo, 10 Sept. to 18 Oct. 1936, *ibid.*, vol. 16, pp. 482–506, passim. Also 23 Aug. 1637, *ibid.*, vol. 17, p. 175.

42. Torricelli to Galileo, 11 Sept. 1632, *ibid.*, vol. 14, pp 387–388.

43. Torricelli to Galileo, 15 March 1641, *ibid.*, vol. 18. p. 308.

44. See Galileo to Castelli, 3 Dec. 1639, and Galileo to Guerrini, 16 and 20 February 1640, *ibid.*, vol. 18, pp. 125–126, 148, and 154–155.

45. Micanzio to Galileo, 14 Oct. 1634, *ibid.*, vol. 16, p. 140.

46. Micanzio to Galileo, 10 Feb. 1635, *ibid.*, vol. 16, p. 209.

47. Michelini to Galileo, 6 Nov. 1638, and 8 Feb. 1639, *ibid.*, vol. 17, pp. 399–400, and vol. 18, pp. 24–25.

48. Leopold to Galileo, 1 March 1640, *ibid.*, vol. 18, p. 165. Galileo's essay addressed to Leopold, *ibid.*, vol. 8, pp. 489–542.

49. Leopold to Galileo, 14 May 1640, Soldani to Galileo, 21 Nov. 1640, *ibid.*, vol. 18, pp. 190 and 274.

50. Settimi to Ferdinand II de' Medici, 14 Dec. 1641, *ibid.*, vol. 18, p. 372.

51. Book One, Section I, *The Mathematical Principles of Natural Philosophy*, trans. by Andrew Motte, revised by Floria Cajori (Berkeley: University of California Press, 1960), pp. 29–39.

52. See my biography of Newton, *Never at Rest* (New York: Cambridge University Press, 1980), pp. 794–795. It will be obvious that my discussion of Newton in this paper draws heavily upon my own book about him, and I will use notes, such as this one, to refer to more extended treatments of divers points.

53. See, for example, the final scholium to Book Two, on the theory of the vortex, *Principles*, pp. 395–396.

54. See especially the passage, "Quaestiones quaedam Philosophicae," in his undergraduate notebook on philosophy, Cambridge University Library, *Add. MS. 3996*, ff. 88–135. The "Quaestiones" are published in full, together with a rendition into twentieth-century English and an extensive commentary, in J. E. McGuire and Martin Tamny, *Certain Philosophical Questions. Newton's Trinity Notebook* (Cambridge: Cambridge University Press, 1983). I discuss the passage in *Never at Rest*, pp. 88–97.

55. *Never at Rest*, p. 587.

56. Christiaan Huygens, *Oeuvres complètes*, pub. by Société hollandaise des sciences (The Hague: Nijhoff, 1888–1950), vol. 9, passim.

57. *Never at Rest*, Chapters 11–15, passim; Frank E. Manuel, *A Portrait of Isaac Newton* (Cambridge, Mass.: Harvard University Press, 1968), Chapter 13, "The Autocrat of Science." I have in mind here also the material in two doctoral dissertations written at Indiana University: Anita Guerrini, *Newtonian Matter Theory, Chemistry, and Medicine, 1690–1713*, (1983); Gerald Funk, *Newton's Clients: Patronage in Science*, (1985).

58. Gregory to Newton, 2 Sept. 1687, *The Correspondence of Isaac Newton*, H. W. Turnbull, et al., eds. (Cambridge: Cambridge University Press, 1959–1977), vol. 2, p. 484.

59. *Philosophical Transactions*, 16 (1686–1687), p. 291.

Descartes and the Art of Persuasion

WILLIAM R. SHEA

Oratory has incomparable powers and beauties . . . poetry
quite ravishing delicacy and sweetness (Descartes, *Discourse on Method*, Part I).

Descartes was acutely sensitive to his public image, and although he considered himself one of the most persuasive writers of his age, he would not have taken kindly to the suggestion that his style was "rhetorical." Although the word did not have in his day the pejorative connotation that it acquired later, it was beginning to be tainted with the implied meaning of "insincere" and "exaggerated." In this respect, it was to be followed a century or so later by another serviceable word, "prosaic," which came to signify "commonplace," "dull," at the very time (such is the irony of history) when much of the creative work in literature was taking place in a prose genre, the novel. I prefer, therefore, to placate Descartes' manes and to speak of his art of persuasion. But before we proceed to examine some of the features of his style, let us briefly survey the fluctuating fortunes of rhetoric.

Part I. Rhetoric: In Good and Ill Repute

Among modern writers, no one tilted his lance at rhetoric with more vigor and panache than Benedetto Croce. In this *Aesthetic*, first published in 1900, Croce argues against the separation of idea and expression, form, and content, and castigates rhetoric for trying to substitute tropes and figures of speech for genuine thought. According to Croce, rhetoric papers over not mere cracks but chasms of ignorance and abysses of absurdity. It is a prime example of superficiality. Nonetheless, "rhetorical categories," he declares, should continue to appear in schools: to be criticized there. The errors of the past must not be forgotten and no more said, and truth cannot be kept alive save by making them combat errors. Unless an account of the rhetorical categories be given, accompanied by a criticism of them, there is a risk of their springing up again.[1]

As Brian Vickers comments in a recent book on the subject, "Rhetoric is to be kept on as a tame Hydra for the budding aesthetician to decapitate from time to time."[2]

Croce's indictment of rhetoric may be strikingly rhetorical, but it has a long and distinguished ancestry. The first and most influential criticism of rhetoric is Plato's *Gorgias* in which Socrates displays great skill in reducing rhetoric to a form of sophistry that enables some people "to appear to know better than the expert."[3] Several centuries later, Immanuel Kant, in his *Critique of Judgement*, was equally dismissive of rhetoric, which he contrasted unfavorably with poetry:

> *Rhetoric* is the art of transacting a serious business of the understanding as if it were a free play of the imagination; *poetry* that of conducting a free play of the imagination as if it were a serious business of the understanding.[4]

A few pages later, Kant adds that rhetoric perverts poetry to its own ends:

> Rhetoric, so far as this is taken to mean the art of persuasion, i.e. the art of deluding by means of a fair semblance . . . is a dialectic, which borrows from poetry only so much as is necessary to win over men's minds to the side of the speaker before they have weighed the matter, and to rob their verdict of its freedom. Hence it can be recommended neither for the bar nor the pulpit.[5]

Plato, Kant, and Croce are weighty opponents, but fortunately rhetoric finds a worthy defender in Aristotle whose *Rhetoric* is a remarkably well-balanced discussion of human communication in general, and a reminder that "all men attempt to discuss statements and to maintain them, to defend themselves and to attack others."[6] According to Aristotle, rhetoric is the study of methods of communication, expression, comprehension, persuasion, and dissuasion. It is a process rather than a self-contained subject, a service industry, capable of meeting a host of needs. Other disciplines, such as ethics, psychology, law, and politics, appeal to rhetoric to organize their technique of invention and presentation. In a sense, it is an armory that is visited by various combatants in need of weapons. That is what Alexander Pope had in mind when, in his *Essay on Criticism*, he referred to Quintilian's *Institutio Oratorica* in these words:

> *In grave Quintilian's copious work we find*
> *The justest Rules, and clearest Method join'd;*

> *Thus useful Arms in Magazines we place,*
> *All rang'd with Order, and dispos'd with Grace,*
> *But less to please the Eye, than arm the Hand,*
> *Still fit for Use, and ready at Command.*[7]

Science and Rhetoric

This encomium of rhetoric will surely have prompted the query, "But wasn't the Scientific Revolution a triumph over Aristotle and all his ways and wiles? Wasn't the birth of science the demise of rhetoric?" Most assuredly, if we are to listen to Thomas Spratt. In his famous *History of the Royal Society*, which appeared in 1667, Spratt deplores that the tools of rhetoric

> are generally chang'd to worse uses: They make the *Fancy* disgust the best things, if they come sound, and unadorn'd; they are in open defiance against *Reason*; professing, not to hold much correspondence with that; but with its Slaves, *the Passions*: they give the mind a motion too changeable, and bewitching, to consist with *right practice*. Who can behold, without indignation, how many mists and uncertainties, these specious *Tropes* and *Figures* have brought on our Knowledge?[8]

This is why the Royal Society reacted so vigorously to this "easie vanity of *fine speaking*" and urged the primacy of *res* over *verba*:

> They have therefore been most rigorous in putting in execution, the only Remedy, that can be found for this *extravagance*: and that has been, a constant Resolution, to reject all the amplifications, digressions, and swellings of style: to return back to the primitive purity, and shortness, when men deliver'd so many *things*, almost in an equal number of *words*. They have exacted from all their members, a close, naked, natural way of speaking; positive expressions; clear senses; a native easiness: bringing all things as near the Mathematical plainness as they can: and preferring the language of Artizans, Countrymen, and Merchants, before that of Wits, or Scholars.[9]

Spratt was echoing Francis Bacon's strictures about a purely verbal science, but he was not himself opposed to using the blandishments of rhetoric. He tells us that when natural philosophy has been cleared of the rank overgrowth of sterile disputation, "The Beautiful Bosom of *Nature* will be Expos'd to our view: we shall enter into its *Garden*, and taste of its *Fruits*, and satisfy ourselves with its *plenty*."[10]

Spratt was not alone in deploring the ill usage of rhetoric. John Wilkins, introducing his *Essay Towards a Real Character and a Philosophical Language* in

1668, spoke of words gobbling up things,[11] and John Locke in his *Essay Concerning Human Understanding* saw the purpose of language as being: "First, to make known one man's Thoughts or ideas to another; Secondly, to do it with as much ease and quickness as possible; and, Thirdly, thereby to convey the knowledge of things."[12] This being the case, rhetoric may be acceptable when we seek pleasure, but it is usually promoted at the expense of truth and knowledge:

> It is evident how much men love to deceive, and be deceived, since rhetorick, that powerful instrument of error and deceit, has its established professors, is publicly taught, and has always been had in great reputation: and I doubt not but it will be thought great boldness, if not brutality in me, to have said thus much against it. Eloquence, like the fair sex, has too prevailing beauties in it, to suffer itself ever to be spoken against.[13]

Seventeenth-century opponents of rhetoric display surprisingly rhetorical skills. They had learned their lesson so well that they ascribed to their native intellect what was really the result of their formal education.

Empty Words or Vacuous Science

The hostility of the English to rhetoric has much to do with Bacon's elegantly phrased dismissal of "ornaments of speech, similitudes, treasury of eloquence, and such like emptiness,"[14] but also with a feeling that the bleak state of science in English universities between 1560 and 1640 was due to the emphasis on rhetoric. This sentiment would linger on throughout the second half of the seventeenth century and the first half of the eighteenth. Indeed, it is instructive to compare the university statutes enacted before that period with those passed under Queen Elizabeth. The Edwardian statutes of 1549 still emphasized mathematics and science, and the article on the method of study stipulated:

> The Student freshly come from a grammar school, mathematics are first to receive. He is to study them a whole year, that is to say, arithmetic, geometry, and as much as he shall be able of astronomy and cosmography. The following year shall teach him dialectics. The third and fourth shall add philosophy.[15]

This article was completely altered in the sets of statutes granted by Queen Elizabeth to Cambridge in 1558 and 1570. It now read: "The first year shall teach rhetoric, the second and third logic, the fourth shall add philosophy."[16] Mordechai Feingold argues in *The Mathematician's Apprenticeship* that science did not fare as badly as the statutes would lead us to believe, but he grants that rhetoric was thrust to the fore until the introduction of the Savilian statutes in Oxford in 1619, and the creation of the Lucasian chair for mathematics in Cambridge almost half a century later in 1663.[17]

Rhetoric was washed onto the shores of sixteenth-century England in the wake of the revival of humanism in fifteenth-century Italy. The recovery of Quintilian's *Institutio Oratoria* in the Renaissance gave rhetoric a new impulse or, better still, a fighting spirit. Whereas Cicero met the objections raised by Plato by suggesting a peace treaty between philosophy and rhetoric in which certain individuals (such as himself) would master both disciplines, Quintilian looked for direct confrontation. He charged that men, real men, engaged in public affairs, and busy with weighty matters of state, "ceased to study moral philosophy, and ethics, thus abandoned by the orators, became the prey of weaker intellects," namely, the philosophers.[18] In the fifteenth century, Lorenzo Valla improved on Quintilian in his dialogue *De Voluptate* where his Christian spokesman rebukes Boethius for being "more friendly with the dialecticians than with the rhetoricians."

> How much better it would have been for him to speak oratorically rather than dialectically! What is more absurd than the procedure of the philosophers? If one word goes wrong, the whole argument is imperilled. The orator makes use of many different procedures: he brings in contrary points, seeks out examples, makes comparisons, and forces even the hidden truth to appear. . . . Boethius ought to have worked in this way; he, like many others, was deceived by excessive love of dialectics. But how much error has been in dialectics, and how no one has ever before written carefully about it, and how it is really a part of rhetoric—about all these things our Lorenzo here has begun to write, very much in accordance with the truth, in my opinion.[19]

Valla's self-commendation may strike us as amusing, but it is part of a serious attempt to subordinate philosophy to rhetoric. In all fairness, it must be said that Valla and other Renaissance writers, such as Juan Luis Vives and M. T. Nizolio, were really campaigning against sterile logic in favor of an art of communication that would be useful for the active and responsible life of a citizen. In their single-minded dedication to the supreme problem of how to live well and eloquently, Renaissance humanists were hostile to forms of intellectual activity that encouraged abstraction. One of the most interesting instances is the sustained attack of Giambattista Vico, for 40 years professor of rhetoric at the University of Naples. In 1700 Vico gave a particularly brilliant public lecture that epitomized his educational ideals and was published as *De nostri temporis studiorum ratione*.[20] Arguing from the primacy of the *vita activa*, and thinking of the man of public affairs, the "committed" citizen, Vico took issue with Cartesian geometry, which he considered detrimental to the training of an open and socially effective mind. In stripping mathematics from any reference to concrete and sensible matter, analytical geometry severed the mind from the realm of analogy and metaphor, the breeding place of plausibility and verisimilitude and, hence, the home of politics. Painting the picture in the blackest possible terms, Vico feared that a training in Cartesian geometry would blind the imagination, enfeeble the memory, slacken the understanding, and destroy perception. [21]

Part II. Descartes' Rhetorical Strategies

Despite Vico's strictures, Descartes does not reduce the art of human communication to geometrical demonstration, and he uses a number of devices that can be called rhetorical in the broad sense of the word. I shall examine a few cases. One of them is belittling, as when he dismisses the French mathematicians, who criticized his geometry, as "two or three flies," or when he describes Roberval as "less than a rational animal," and Claude Petit as "a little dog." [22] Other instances of this unprepossessing method are his description of Jean de Beaugrand's letters as suitable for "toilet paper," and Fermat's work as "plain shit," a word hardly softened for being in Latin. [23]

This overbearing and arrogant style was not exclusive to Descartes. Compare, for instance, the following passage from his letter to Mersenne: "I would gladly have them believe that if I am wrong about the motion of the heart, refraction, or anything else that I have discussed in more than a couple of lines, then the rest of my philosophy is worthless," [24] with Galileo's statement in a postil to a book by the Jesuit Orazio Grassi (who wrote under the pseudonym of Sarsi): "What do you want, Mr. Sarsi, if it was given to me alone, and to no other, to make all the new discoveries in the heavens?" [25]

The age of Galileo and Descartes had long ceased to be an age of chivalry and had become the age of Spanish pride and Baroque punctilio. It is symptomatic that in 1636, the year before the publication of the *Discourse on Method*, Paris acclaimed Corneille's immensely successful play, *Le Cid*, which revolves entirely on a point of honor. It was also an age when it was permitted, on occasion, to appear with a mask. In a cryptic passage from his posthumously published notebook, Descartes wrote:

> Actors, taught not to let any embarrassment show on their face, put on a mask. I shall do the same. So far, I have been a spectator in this theatre which is the world, but I am now about to mount the stage, and I come forward masked. [26]

Descartes can be said to have ascended the stage with the publication of his first book, the *Discourse on Method*, in 1637 when he was already 41 years old.

This work contains a famous description of how he came to discover his new method of philosophizing:

> At that time I was in Germany, where I had been drawn by the wars which are not yet at an end. And as I was returning from the coronation of the Emperor to join the army, the setting in of winter detained me in a quarter where, finding no society to divert me and fortunately having no cares or passions to trouble me, I remained the whole day shut up alone in a stove-heated room, where I had complete leisure to occupy myself with my own thoughts.[27]

"Finding no society to divert me and fortunately having no cares or passions to trouble me . . . I had complete leisure to occupy myself with my own thoughts." How detached and serenely philosophical all this sounds! Descartes' description of his state of mind would have us believe that he arrived at his insight in the attitude that a sculptor would select if asked to represent "the thinker." It is, in fact, a posture. Some 20 years earlier, Descartes had recorded the incident in language far removed from the cool and calm prose of the *Discourse on Method*. It is couched in the language of dreams, and although the manuscript in which he wrote out a detailed account of his visionary experience has vanished, it was seen by Leibniz during his visit to Paris in 1675–1676, and was translated by Adrien Baillet in his biography of Descartes published in 1691.

A Land of Dreams

In this autobiographical manuscript, Descartes tells us that on the night of November 10–11, 1619 he had in rapid succession not one but three dreams, "which he imagined could only come from on high." In the first dream, he was frightened by ghosts and buffeted by a strong wind that kept him from advancing to where he wished to go. He felt, for instance, a weakness on the right side; a whirlwind made him spin three or four times on his left foot; others were straight and steady while he wavered; and so forth. Descartes woke up in a fright, confessed his sins to the Almighty, and fell asleep again. The second dream ended with a piercing noise like a clap of thunder, which awoke him. On opening his eyes,

> he perceived a large number of fiery sparks all around him in the room. This had often happened to him at other times; it was nothing extraordinary for him to wake in the middle of the night and find his eyes sparkling to such a degree as to give him glimpses of the objects nearest to him.[28]

Quite a feat, even for a philosopher!

There can be no doubt that Descartes believed that he could see in the dark. He states in the *Optics* that light streams out of the eyes of cats, and that this is also possible for humans who rise above the ordinary—namely, people like himself. [29] But it is unlikely that he would have believed in his own prowess if he had not read about its possibility elsewhere, and I doubt whether he would have committed it to writing if he had not been less certain of its rhetorical efficacy. In an age that was fond of anecdotes and raised on the classics, Descartes could rely on many readers remembering that the Emperor Tiberius could see in the dark. [30]

The third dream, which followed fast upon the second, was peaceful by contrast. He saw a book that he took for a dictionary, and a collection of poems, which he opened at random. He fell on a poem by Ausonius that began " *Quod vitae sectabor iter?*" At that moment, an unknown person handed him another poem that began with "*Est & Non.*" It then occurred to Descartes, in the midst of his dream, to ask himself whether he was dreaming. He not only concluded that he was dreaming, but he started to interpret the dream while still asleep. When he awoke, "he continued the interpretation of the dream on the same lines." Note the continuity between the sleeping and the waking state. He judged the dictionary to be the "sciences gathered together," the collection of poems "the union of philosophy and wisdom," and the poets assembled in the collection "revelation and inspiration, by which he hoped to see himself favored." The "*Est & Non*" of the poem, he interpreted as "the Yes and No of Pythagoras," meaning truth and error in human knowledge and the secular sciences. The clap of thunder that he heard in the second dream, he took to be "the signal of the Spirit of truth descending to take possession of him." Lest this new Pentecost be greeted like the first one with jeers ("They are filled with new wine," *Acts* 2, 13), Descartes, like Peter before him, affirms that he was not drunk, having "passed the evening and the whole day in a condition of complete sobriety," and for that matter, not having touched wine for three months!

I do not believe that Descartes had forgotten all this when he wrote the *Discourse on Method*. Granted that we recollect what we want to remember and that we cut and edit our life history, we cannot seriously entertain the notion that from the rationalist standpoint he had reached in 1637 the past was so distorted that a breakdown occurred somewhere along the memory line. Descartes was not the kind of person psychoanalysts love to see lying on their couch, and more cogently, he kept the Dream Manuscript with him for 30 years.

I believe it is much more profitable to raise the question whether Descartes deliberately chose the dream as a literary device that allowed the use of symbols that would appear incongruous if the narrator were in the waking state. Every educated person in the seventeenth century was familiar with the Dream of Scipio and knew from both biblical and classical sources that God communicates with men in dreams. As a poetico-philosophical device, it was not uncommon in the sixteenth and seventeenth century.[31] For instance, in Rodophilus Staurophorus' *Raptus Philosophicus*, published in German in 1619, we find an account of the dream of a young man at the crossroads. He wonders what path to take and—one readily guesses—chooses the straight and narrow. After several dangerous incidents, he meets a young woman who asks him, "Where are you going? What Spirit brings you here?," and shows him a book "that contained all that is in earth and in heaven but not ordered methodically." A young man dressed in white then reveals to him that this woman "is Nature . . . at the present time unknown to scientists and philosophers."[32]

Henri Gouhier believes that Descartes may have been acquainted with this work, but that its influence was "*purement* ornamentale."[33] But there is more than pure ornament here, since Baillet explicitly states that Descartes had been *expecting* significant dreams for some time, and that "the human mind had no share in them." He writes:

> the Genius which had been exciting in him the enthusiasm with which,
> as he felt, his brain had been inflamed for several days had predicted
> these dreams to him before he retired to rest.[34]

Our acquaintance with the vast Renaissance literature on dreams is too slight to allow us to see Descartes' account against the background that he shared with his contemporaries and that was probably so familiar as not to be considered worth mentioning. Francesco Trevisani has drawn attention to Cardano's symbolic interpretations of dreams, and there are scores of such works to be explored.[35] In 1599 Giovanni Battista Nazari published his *Della tramutatione metallica sogni tre* in which a rustling of weeds calls out, "*Quo viator iter, tu avaritia dementis.*"[36] A famous contemporary of Descartes, J. B. van Helmont, gives 1610 as the date of his dream of enlightenment:

> In the year 1610 after a long weariness of contemplation, that I might
> acquire some gradual knowledge of my own minde, since I was then
> of opinion, that self-cognition was the Complement of wisdom, fallen
> by chance into a calm sleep, and wrapt beyond the limits of reason, I
> seemed to be in a Hall sufficiently obscure. On my left hand was a
> table, and on it a faire large Vial, wherein was a small quantity of
> Liquor: and a voice from that liquor spake unto me: "Wilt thou Honour
> and Riches?"[37]

Descartes believed in the virtues of rhetoric. He also believed in inner voices. To Princess Elizabeth, with whom he entertained a free and easy correspondence, he wrote as late as 1646:

I even make bold to believe that inner joy has a secret force to make Fortune more favourable. I would not write this to people with weak minds lest they should be led to superstition, but, in the case of your Highness, I rather fear that she will laugh at my credulity. But I have a very large number of experiences, as well as the authority of Socrates, to confirm my opinion. My experience is such that I have often noticed that what I undertake gladly and without an interior repugnance generally goes well even in games of chance where Fortune alone holds the sway.[38]

A World of Fiction

If we now leave the realm of the philosophical visionary and enquire about the work of the scientist, we find that Descartes had to avail himself of the resources of literary fiction and rhetoric in order to describe his new cosmology. The problem was to get people to listen and to walk in his footsteps without pointless preambles and justifications. Descartes believed that if he could avoid giving offence, he would easily persuade an unbiased reader that he was right. How to do this worried him for several months in 1630 until he struck upon the idea of writing what he terms "a fable," and what we would call a piece of science fiction. This is how Descartes came to ask his readers to allow their thought "to leave this world for a wholly new one that I shall cause to rise before you in imaginary spaces."[39]

The interaction between science and rhetoric begins here.

In choosing to describe a new world from the moment of its creation to its present state, Descartes realized that he was entering a difficult and dangerous territory. How the world began was described in the Bible, but this was not open to the investigation of natural philosophers. Sir Robert Boyle, for instance, carefully distinguished between the origin of the universe (which he left to the theologians), and its continuing course (which he claimed for himself and his fellow scientists). [40] Now Descartes accounted for the origin of the universe by assuming that matter arranged itself according to the laws of motion. This appeared preposterous to many of his adversaries, and to Leibniz it later appeared downright impious. He described Article 47 of the Third Part of Descartes' *Principles of Philosophy* as "the Πρωτὸν ψευδος and the basis of philosophical atheism."[40] The incriminated article runs in part as follows:

> These few assumptions seem to me sufficient to serve as the causes from which all the effects observed in our universe arise in accordance with the laws of nature set out above. And I do not think it is possible to conceive of alternative principles for explaining the real world that are simpler, or easier to understand, or even more probable. . . . In fact

it makes very little difference what initial suppositions are made, since all subsequent change must occur in accordance with the laws of nature. And there is scarcely any supposition that does not allow the same effects (albeit more laboriously) to be deduced in accordance with the same laws of nature. For by the operation of these laws matter must successively assume all the forms of which it is capable; and, if we consider these forms in order, we will eventually be able to arrive at the form which characterizes the universe in its present state. [42]

It appeared to Leibniz that if matter can successively receive all possible forms, then whatever we can conceive (however extravagant, absurd, or dishonest) has either already occurred or will occur in the future. This entails that everything that can happen necessarily happens, and that there is no Providence and no free choice. But lest we suspect him of prejudice against Descartes, Leibniz ends his criticism with the most rhetorical of disclaimers:

> To express in a few words the feelings I have for an author whose reputation I am unjustly accused of trying to ruin (an enterprise that would be as unfair as it is impossible), let me say that anyone who fails to recognize the great talent of Descartes cannot be very sharp, and that anyone who knows and appreciates only Descartes and his followers will never achieve much. [43]

Before ascribing these sentiments to a personal *animus* against Descartes, we should note that Newton experienced the same *vertigo* at the thought of unbridled cosmological speculation:

> it's unphilosophical to seek for another origin of the world, or to pretend that it might arise out of a chaos by the mere Laws of Nature; though being once form'd, it may continue by those Laws for many Ages. [44]

"Let There Be Light"

In a Christian society in which the biblical account of creation was still normative, a cosmological novel avoided a head-on collision. Nonetheless, Descartes would have been happy if some kind of concordance had been forthcoming, and he made an attempt in this direction. The problem was twofold: the first concerns the *order* of creation, the second the *state of perfection* in which the creatures appear in the book of Genesis.

God's first words in the Bible are, "Let there be light," and Descartes was justifiably pleased at the centrality of the notion of light in his system. But in *Genesis*, God creates light before the firmament and the heavenly bodies, whereas in Descartes' *World*, light is the result of the action of celestial bodies. Descartes was aware of the discrepancy as early as 1630 when he wrote to Mersenne:

> I am now ordering the chaos to produce light. This is one of the highest and most difficult tasks that I could face because it involves virtually the whole of physics. I have to think of a thousand different things at the same time in order to find a way of stating the truth without shocking someone or offending received opinions. I want to take a month or two to think of nothing else.[45]

The outcome was a rough concordance: (a) in the first verse of the Bible, God creates the heavens and the earth; in the Cartesian *World*, He first creates matter; (b) in the Bible, the earth is "formless" (*Gen.* 1, 2); in *The World* it is "a chaos;" (c) all things are ordered by God in the first chapter of *Genesis*; in the *World*, He "establishes everything in number, weight, and measure."[46] The fit was less than perfect, but when Descartes decided to publish a revised version of *The World* in 1641, he thought that it was so good that he wrote to Mersenne that his *Principles of Philosophy* would contain an explanation of the first chapter of *Genesis*. It is probably at this time that he confided to an unidentified correspondent:

> I am not moving fast, but I am moving. I am now describing the origins of the world in which I hope to include most of my physics. Rereading the first chapter of the book of *Genesis*, I was astonished to discover that it can be completely explained on my view, and much better, it seems to me, than on any other view. I had never dared to hope for so much before, and I am now resolved, after giving an account of my new philosophy, to clearly show how all the truths of Faith agree much better with my philosophy than with that of Aristotle.[47]

Descartes even made an attempt to learn Hebrew, although he did not get very far. In a seventeenth-century memoir, Descartes is said to have called on Anna-Maria Van Schuurman (1607–1678), a prodigy who knew most European languages, including Latin and Greek, as well as Syriac, Chaldean, Arabic, and Turkish. On this particular morning, around 1640, Descartes found her reading the Scriptures in Hebrew. He expressed surprise that she should be wasting her time "on such a trifling matter." When Miss Schuurman remonstrated and tried to show him that the word of God should be read in the original, he replied that he had once had the same notion and had begun "to read the first chapter of *Genesis* on the creation of the world, but think about it as he might, he could not find anything clear and distinct, nothing that he could grasp *clare et distincte*."[48]

In 1648 Descartes was visited at Egmond by a 20-year-old admirer named
Frans Burman. He entertained the young man at dinner and gave frank and lively
answers to his questions. One touched on the topic of concordance, and Descartes
replied that he had tried to fit his account to the story of *Genesis* but had decided to
give it up and leave the whole thing to the theologians, since the correct interpretation might well be metaphorical, as would seem to be the case with the six days
of creation.[49]

Perfection from the Start

Whatever the sequence of events in the history of creation, the Bible clearly
affirms that things were created in their state of perfection. Descartes had no wish
to see his opponents accuse him of following Lucretius and the atomists rather than
Moses, and in the *Discourse on Method*, he clearly states that the world was created as we actually find it. In the *Principles of Philosophy*, he is even more explicit.
The sun, the stars, and the living plants appeared in all their perfection. Adam and
Eve were created as adults. But how can we reason *genetically* if there is no real
development and things are created in their full perfection?

Descartes' answer lies in his conviction that the laws that govern the present
world are those of any conceivable world. While there may be an infinite number
of possible states for a world made of matter and motion, there is only one state of
equilibrium: the one we witness. The text of the *Discourse* is explicit:

> It is certain, and it is an opinion commonly accepted among theolo-
> gians, that the act by which God now preserves it is just the same as
> that by which he created it. So even if in the beginning God had given
> the world only the form of a chaos, provided that he established the
> laws of nature and then lent his concurrence to enable nature to operate
> as it normally does, we may believe, without calling the miracle of
> creation in question, that by this means alone all purely material things
> could, in the course of time, have come to be just as we now see them.
> And their nature is much easier to conceive if we see them develop
> gradually in this way than if we consider them only in their completed
> form.[50]

Adam and Eve were created as adults, for such is "the doctrine of our
Christian faith, and our natural reason convinces us that it was so." Given the
infinite power of God, we cannot imagine that he ever created anything that was
not wholly perfect of its kind. "Nevertheless," continues Descartes,

> if we want to understand the nature of plants or of men, it is much
> better to consider how they can gradually grow from seeds than to

consider how they were created by God at the very beginning of the world. Thus we may be able to think up certain very simple and easily known principles which can serve, as it were, as the seeds from which we can demonstrate that the stars, the earth and indeed everything we observe in this visible world could have sprung. For although we know for sure that they never did arise in this way, we shall be able to provide a much better explanation of their nature by this method than if we merely described them as they now are.[51]

This passage raises the whole problem of Descartes' rhetoric. Was he merely seeking a way of reassuring the censors that he did not intend to deviate from the straight and narrow path of orthodoxy? Or was he trying to persuade himself as well as others that he could have it both ways, namely, that he could argue *genetically*, although the world had really appeared in full array at the very instant of the divine fiat? I believe that Descartes convinced himself that his developmental approach was methodologically sound even in the static universe that seemed to be required by the Catholic faith to which he subscribed. Indeed, rhetoric triumphed to the point of leading him to formulate the principle that appears as the Third Rule of his famous Four Rules of Reasoning:

to direct my thoughts in an orderly manner, by beginning with the simplest and most easily known objects in order to ascend little by little, step by step, to knowledge of the most complex, and by *supposing some order even among objects that have no natural order of precedence.*[52]

In the rhetorical perspective that we have been examining, this rule makes sense. But for those who, like Leibniz, sought to understand it as a straightforward general rule for philosophizing it seemed perfectly inane.[53] This may indicate that the greatest French rationalist of the seventeenth century cannot be understood without the aid of rhetoric.

NOTES

1. Benedetto Croce, *Aesthetic*, translated by D. Ainslie, 2nd edition (London: Longman, 1922), pp. 72–73.

2. Brian Vickers, *In Defence of Rhetoric* (London: Clarendon Press, 1988), p. 206. This is an excellent survey of the history of rhetoric in English. See also S. Ijsseling, *Rhetoric and Philosophy in Conflict. An Historical Survey* (The Hague: M. Nijhoff, 1976).

3. Plato, *Gorgias*, 459d, translated by W. D. Woodhead, in Plato, *The Collected Dia-*

logues, Edith Hamilton and Huntingdon Cairns, eds. (Princeton: Princeton University Press, 1980), p. 242.

4. Immanuel Kant, *The Critique of Judgement*, translated by J. C. Meredith (Oxford: Oxford University Press, 1928), p. 185.

5. *Ibid.*, p. 192.

6. Aristotle, *Rhetoric*, 1355a 4–6, translated by W. Rhys Roberts in Aristotle, *The Complete Works*, Jonathan Barnes, ed., 2 vols. (Princeton: Princeton University Press, 1984), vol. 2, p. 2152.

7. Alexander Pope, *Essay on Criticism*, lines 669–674.

8. Thomas Spratt, *History of the Royal Society* (London, 1667), Jackson I. Cope and Harold Whitmore Jones, eds. (facsimile edition, St. Louis: Washington University Press, 1958), p. 112.

9. *Ibid.*, p. 113.

10. *Ibid.*, p. 327.

11. John Wilkins, *An Essay Towards a Real Character and a Philosophical Language* (London, 1668), p. 18, quoted in Brian Vickers, "The Royal Society and English Prose Style: A Reassessment," in Brian Vickers and Nancy S. Struever, *Rhetoric and the Pursuit of Truth* (Los Angeles: William Andrew Clark Memorial Library, 1985), pp. 6, 65, n. 7.

12. John Locke, *Essay Concerning Human Understanding*, Alexander Campbell Fraser, ed., 2 vols. (London, m.d., reprint, New York: Dover, 1959), Book 3, Chapter 10, vol. 2, p. 142.

13. *Ibid.*, pp. 146–147.

14. Francis Bacon, "Parasceve, or Preparative towards a Natural and Experimental History," in *Works*, James Spedding, Robert Leslie Ellis, et al., eds., 14 vols. (London: Longman, 1857–1874), vol. 4, p. 254.

15. *Collection of Statutes for the University and Colleges of Cambridge*, translated by J. Heywood (London, 1840), p. 1, quoted in Mordechai Feingold, *The Mathematician's Apprenticeship: Science, Universities and Society in England 1560–1640* (Cambridge: Cambridge University Press, 1984), p. 24.

16. *Ibid.*, pp. 24–25. Feingold's book is an invaluable source of information about the contribution of English universities to scientific activity.

17. *Ibid.*, pp. 1–22. Charles Webster considers that science remained marginal in both Oxford and Cambridge until 1640. See *The Great Instauration: Science, Medicine and Reform 1626–1668* (London: Duckworth, 1975), pp. 115–129.

18. Quintilian, *Institutio Oratorica*, translated by H. E. Butler, 4 vols., Loeb Classical Library (London: Heinemann, 1986), Book I, Pr. 13–14, vol. I, p. 13.

19. Lorenzo Valla, *On Pleasure. De Voluptate*, translated by Maristella Lorch and A. K. Hieatt, eds. (New York: Abaris Books, 1977), p. 273.

20. This important work received little attention in the English-speaking world until recently when it was translated by Elio Gianturco: Giambattista Vico, *On the Study Methods of Our Time* (Indianapolis: Bobbs-Merrill, 1965). A good study of Vico and rhetoric is now available in English. See Michael Mooney, *Vico in the Tradition of Rhetoric* (Princeton: Princeton University Press, 1985). An excellent collection of general essays on Vico is Paolo Rossi, *Le sterminate antichità. Studi vichiani* (Pisa: Nistri-Lischi, 1969).

21. Girambattista Vico, *On the Study Methods of Our Time*, pp. 21–24.

22. René Descartes, *Oeuvres*, C. Adam and P. Tannery, eds., 11 vols. (Paris: Cerf, 1897–1913; reprinted with additions, Paris: Vrin, 1964–1979), vol. II, pp. 671, 190, 267, 533.

23. *Ibid.*, vol. III, p. 437, vol. II, p. 464.

24. *Ibid.*, vol. II, p. 501 (letter of 9 February 1639).

25. Galileo Galilei, *Opere*, A. Favaro, ed., 20 vols. (Florence: Barbèra, 1890–1909), vol. VI, p. 383, n. 13.

26. Descartes, *Cogitationes Privatae*, in *Oeuvres*, vol. X, p. 213. English translation by John Cottingham in *The Philosophical Writings of Descartes*, 2 vols. (Cambridge: Cambridge University Press, 1985), vol. I, p. 2.

27. Descartes, *Discourse on Method*, Part II in *Oeuvres*, vol. VI, p. 11.

28. Descartes, *Olympica*, in *Oeuvres*, vol. X, p. 182. The source is Antoine Baillet, *La Vie de Monsieur Des-Cartes*, 2 vols. (Paris, Daniel Horthemels, 1961; facsimile edition, Geneva: Slatkine Reprints, 1970), vol. I, pp. 80–86.

29. Descartes, *Optics [La Dioptrique]*, in *Oeuvres*, vol. VI, p. 86.

30. The story is told by Sextus Empiricus, *Outline of Pyrrhonism*, Book I, 84, in *Sextus Empiricus*, translated by R. G. Bury, 4 vols., Loeb Classical Library (London: Heinemann, 1933–1949), vol. 1, p. 51.

31. See André Chastel, *Marsile Ficin et l'art* (Geneva: Droz, 1954), pp. 147–148, and Jean-Marie Wagner, "Esquisse du cadre divinatoire des songes de Descartes," *Baroque* 6 (1973), pp. 81–95. In Cicero's widely read *De divinatione*, Quintus, Cicero's interlocutor, remarks: "The Stoic view of divination smacked too much of superstition. I was more impressed by the reasoning of the Peripatetics, of opinion there is within the human soul some sort of power—'oracular' I might call it—by which the future is foreseen when the soul is inspired by a divine frenzy, or when it is released by sleep and is free to move at will" (Cicero, *De Senectute. De Amicitia. De Divinatione*, translated by W. A. Falconer (London: Heinemann, 1923), p. 483).

32. Quoted in P. Arnold, *La Rose-Croix et ses rapports avec la Franc-Maçonnerie* (Paris: G.-P. Larose, 1970), pp. 160–161.

33. Henri Gouhier, *Les première pensées de Descartes* (Paris: Vrin, 1958), p. 140.

34. Descartes, *Oeuvres*, vol. X, p. 186. See my *The Magic of Numbers and Motion: the Scientific Career of René Descartes* (Canton, Mass.: Science History Publications, 1991), pp. 93–120.

35. Francesco Trevisani, "Symbolisme et interprétation chez Descartes et Cardan," *Rivista Critica di Storia della Filosofia* 30 (1975), pp. 27–47.

36. Gio. Battista Nazari, *Della tramutazione metallica sogni tre* (Brescia: Pietro Maria Marchetti, 1599), p. 10.

37. J. B. Van Helmont, *Ternary of Paradoxes*, translated by Walter Charleton (London, 1650), p. 123.

38. Letter of November 1646, in *Oeuvres*, vol. 4, p. 529.

39. Descartes, *The World* [*Le Monde*], in *Oeuvres*, vol. X, p. 31. This work is an early version of the *Principles of Philosophy* that he published in 1644.

40. Robert Boyle, *About the Excellency and Grounds of the Mechanical Hypothesis*, in *Works* (London, 1772), vol. IV, pp. 68–69. For an excellent discussion, see Paolo Rossi, *I Ragni e le Formiche* (Bologna: Il Mulino, 1986), pp. 28–42.

41. G. W. Leibniz, Letter of November 1680 to Christian Philipp, in *Die Philosophischen Schriften*, C. J. Gerhardt, ed., 7 vols. (Berlin, 1875–1890; reprinted, Hildesheim: Georg Olms, 1960–1965), vol. IV, p. 283.

42. Descartes, *Principles of Philosophy*, Part III, Art. 47, in *Oeuvres*, vol. VIII–1, pp. 101–103.

43. Leibniz, *Die Philosophischen Schriften*, vol. IV, pp. 341–342.

44. Isaac Newton, *Opticks* (London, 1721; reprinted, New York: Dover, 1952), p. 402.

45. Letter of 23 December 1630, in *Oeuvres*, vol. I, p. 194.

46. Descartes, *The World*, in *Oeuvres*, vol. XI, p. 47. This is itself a quotation from the *Book of Wisdom*, 11:21.

47. Descartes, *Oeuvres*, vol. IV, p. 698. The date of this letter is uncertain. Etienne Gilson makes a strong case for assigning it to 1641 (René Descartes, *Discours de la méthode*, Etienne Gilson, ed., 4th edition (Paris: Vrin, 1967), pp. 381–382).

48 The anecdote appeared in the anonymous *Vie de Jean Labadie* (Paris, 1670), quoted in *Oeuvres*, vol. IV, pp. 700–701.

49. Conversation with Burman, 16 April 1648, in *Oeuvres*, vol. V, pp. 168–169.

50. Descartes, *Discourse on Method*, Part 5, in *Oeuvres*, vol. VI, p. 45.

51. Descartes, *Principles of Philosophy*, Part I, Art. 45, in *Oeuvres*, vol. VIII–1, p. 100.

52. Descartes, *Discourse on Method*, Part 2, in *Oeuvres*, vol. VI, pp 18–19.

53. Leibniz, *Die Philosophischen Schriften*, vol. IV, p. 331.

The Person-Centered Rhetoric of Seventeenth-Century Science[*]

PETER MACHAMER

F rom the eighteenth century onwards, much has been written attempting to analyze the nature of seventeenth-century science. Sometimes this has appeared under the guise of trying to show the nature of the Scientific Revolution; sometimes in broader scope, trying to account for the Age of Genius. As we know, it was not just science that changed during this period, but philosophy, conceptions of the nature of government, economic values, social relations, and religion and theology. Also very clearly, but less intelligibly, the very idea of the human being and the powers of being human became a great cause for reflecting and theorizing. What I wish to argue in this essay is that the seventeenth century was fundamentally and importantly an ego-centered age. If you will, it was a neo-Protagorean age in which epistemological and constructive versions of the doctrine "man as the measure" are found in all aspects of thought. For clarity though, the seventeenth-century person perspective was not Protagoras' person as perceiver as the measure, but rather person as rational.

This doctrine is manifest in the language of seventeenth-century books. The new concerns brought with them different ways of writing, changed emphases and conceptions about what was important. These differences appeared in the form of

---

[*]My thanks to Barbara Boylan and Merrilee Salmon for their work in trying to make this paper intelligible. Special acknowledgment is due to Ted McGuire who convinced me that the Protagorean metaphor was a good way of telling my story and to Marcello Pera who made me see more clearly how it was a neo-Protagorean principle.

increased use of person-referring locutions: anthropomorphic models, metaphors and descriptions; explicit physiological and psychological discussions; passages dealing with the role of persons in the world and in the cosmological scheme, and above all in great attention explicitly or implicitly paid to the role of human reason and emotions. The rhetoric of the seventeenth century gave primacy to terms describing relativity, subjectivity, and interpretation. It emphasized the ego—not only the *ego cogito* of thought, but also the *ego amo* of affective human emotions. It was person centered or focused. This new discourse brought out the problematic character of objective, absolute, and certain knowledge. It drew attention to the constructive and cognitive role of the observer, the knower. It was perspectivalist, interpretive, and probabilistic from a human point of view. These new rhetorical turns were constituative of the new ways of thinking and writing; they were not meant just as persuasive or ornamental devices. In this regard the rhetoric that appeared in seventeenth-century scientific texts and in philosophical works that dealt with science was very different from what preceded it and different again from what followed in the eighteenth century.

Earlier Renaissance traditions had focused on the individual human being and developed changes in how people thought about themselves and their social and physical world, including a crisis about how to think of the nature of knowledge (*scientia*). Yet despite this person-centered concern, the objectivity and certainty of science and knowledge remained a goal. Increasingly, however, the basis for objective knowledge are called into question. The unsettling character of the times brought attendant scepticism and worries about what could be objective, what could be certain, and what by nature had to be probabilistic or only morally certain. This change in patterns of thought was exhibited in talk about the constructive nature of geometry, the relativity of motion, the relational construals of space and time, in subjective concerns about the individual as a first-person knower, the nature of reason and method, and in the function of the individual's social and political roles. Tellingly, these strands all came together in the seventeenth century in the reliance on the human's power to experience, to reason, and to act in ways that made the human mind the measure of all things or the standard by which all things were to be judged. But the tension of the time was how could a person-centered, human perspective serve as a basis for anything absolute, certain, and objective?

The seventeenth-century story of the new Protagorean perspective is a story of natures: the nature of the physical world, including the nature of motion; the nature of the state or commonwealth; and the nature of God. Though the nature of the human being was most important, it has to be clear that the epistemological aspects of this story that emphasized a person's subjective, inner experience (the way of ideas, as it is sometimes called) were only one aspect of this ego-centered way of thought. The human as knower and maker was the lens through which most of the other natures were seen. How through knowledge and action did the human fit into the physical world, the social world, and the transcendent world of God? In short, what was the place of the human being in nature, in the world?

The perceived inadequacy and constant criticism of older Aristotelian and Scholastic views opened new ways for theorizing about what natures really were. The seventeenth-century thinkers could not cast off completely the old rhetoric deriving from Aristotle; however, their goals were anything but Aristotelian. Most writers kept Aristotle's talk about natures and proper forms while interpreting these terms in ways that were very different. Though similarities in the terms and syntax remained, seventeenth-century writers found a need to rethink, usually in a systematic way, the concepts that underlay the principles or laws (*archai*) and the active forces or powers (*energia*). In a way this was a causal question about how best to describe the states and changes in the world, which included questions about how to use mathematics. In this attempt, of course, our seventeenth-century heroes borrowed much from the past. Galileo chose Archimedes to build upon, while Descartes kept much from Suarez, and many others revived certain strands of atomism, scepticism, and neo-Platonism. But the basic theme throughout concerned the powers of a human being. How did the human being experience the world and react to it? What in nature grounded a human's knowledge about the world, the state, and God? And how could these relations be described, so that these thinkers could come to understand not only the nature of the different sorts of entities in the world, but also the very nature of causality itself? The goal was nothing less than a series of attempts to completely restructure our knowledge of the nature of things from an egocentric, experiential point of view.

The question I wish to reflect upon concerns how the human, epistemological underpinnings of this wave of thought had important influences upon the "new science." It is my thesis that as a result of new thinking about the nature of human beings, and especially their nature as experiencing and reasoning subjects, natural science took a new direction in the seventeenth century. It was a direction that placed the person and the person's subjective, very personal feelings and thoughts at the heart of the early seventeenth-century theorizing. My second thesis is that this radical person-centered period comes to an end in science, but not in philosophy, political theory, or theology, because of the influence of Newton's *Principia*. Newton, by ignoring the philosophical tradition that had grown up before and around him, was successfully able to ignore certain epistemological concerns. Newton believed that he had shown how science could be "objective" again, and thus could ignore the personal element.

Obviously this is a long and detailed story. I can only sketch some of the central elements here, and I will deal only with familiar thinkers. By pointing out the person-centered theme in the "great" men and briefly showing some crucial instances of how constructivity, subjectivity, and relativity played roles in their thought, I hope to hint how the perspectivalist thesis could explain "lesser" figures and more abstract social and intellectual phenomena. More specifically, my thesis about the neo-Protagorean program explains why so many philosophers and scientists were concerned in naturalistic ways with the human passions and emotions (i.e., the often neglected and unexplained fact that so many seventeenth-century thinkers wrote treatises on the passions); explains why probability and scepticism played such an important role during the century; explains how the Protestant idea about individual relations with God fit in with more general ideas about philosophy and science; explains why determinism and free will became a big issue of the age; explains why a new theory of the state, personal property, and economic wealth emerged in the person-based form that it did; and explains why voluntarism and natural theology, *prima facie* antithetical theological positions, tended to grow up along side each other and were often held by one and the same thinker. Similarly my concern with the knowing subject explains why the treatments of method and reason were so important, why experimentalism and mathematics in science were dominant themes during the century and yet were not opposed to one another in most peoples' thought, and why scientific knowledge became the standard and model for rational thought. Not a meager set of claims for a short essay! I believe that much of this story could be told from a social point of view, but I shall not do so here.

Galileo

Let me start, as is fashionable, with Galileo. Here you will see my *modus operandi*. I will draw your attention to a bit of Galileo's work that is commonplace, then interpret it from the person-centered perspective—in a way that shows my general theme of constructivity, relativity, and subjectivity. I will adduce some supporting evidence from other areas of Galileo's thought and then move onto another thinker.

Galileo was not a philosophical or deeply systematic thinker, so we should not expect to find his uses of the neo-Protagorean principle well developed or highly stressed. Yet they exist in many areas of his work.

Galileo was concerned throughout his life with the nature of motion. In his early work, *De Motu* (1590), he tried to show that the analysis of all motion could be reduced to problems of the simple Archimedian machines, viz., to show how problems about floating bodies reduce to problems of balances, and then to treat in the same way the motion of pendula, movements along inclined planes, and natural motion including free fall. I would argue, but will not here, that this goal essentially remained in place through the publication of *Discorsi* in 1638.

This use of the Archimedian machines as a static model for handling problems of motion remained dominant throughout the seventeenth century. Indeed, I claim that it was this static model that came to be shared by Descartes, Hobbes, Hooke, Wren, Wallace, Huygens, and most everyone, until it was replaced by "Newtonian" dynamics. The latter really didn't come until after Newton with the acceptance of algebraic ways of representing motion.

The statical tradition, with which Galileo began, used the theory of geometrical proportions to deal with motion mathematically. Proportional geometry, particularly in its Archemidian form, was taken to be the mandatory way of representing and solving problems about the nature of motion.

Proportional geometry is inherently comparative and relational. It measures one thing by showing its relation to another, which is conceived of as a standard. As such, proportional geometry only makes claims relative to some other basis of knowledge or accepted standard. There is no way within this type of mathematics to arrive at absolute numerical values. The ideas of objectivity and certainty as being absolute measures were not part of this geometrical picture. Measures had to be based upon standards assigned from outside or else be strictly comparative and relational. Assignments of standards by which to assign numbers in his proofs were arbitrary for Galileo. The measure for the proof was thus given by man.

Yet geometry in its more Euclidian guise with its axiomatic structure also became the chief model for objective proof or proper method. That seventeenth-century geometry was both proportional and axiomatic is not now, nor was it then, often distinguished. Often one is left with the impression that the aura of certainty that attached to the aximotic proof structure was carried over to the character of relational proofs just because they were both "geometrical."

Thinking about motion in the relational terms of proportional geometry made it easy to think about the relativity of motion. The measure of a motion too is always in terms of something else taken as a standard. In this way the relativity of motion, as well as this constant use of proportional geometry, was tied with a neo-Protagorean way of looking at the world. In fact the rhetoric of the relativity of motion often was illustrated and made intelligible through the use of person-centered examples, as in Galileo's introduction of the topic through a person noticing that the moon follows him. But it was person centered with a difference, as we shall see. In later times people felt that only when a nonrelational mathematics, algebra, was articulated did science achieve a proper standard of objectivity.

In *Dialogo* (1632) Galileo explicitly introduced relativity into his treatment of natural and impressed motions on a moving earth. The principle of the relativity of observed motion had as its example a man writing a letter on a ship as it travels. The use of the ship example to show relativity of motion was not new to Galileo, though his emphasis on the perceptual way of treating the example and the principle was unusual.

The principle of the relativity of perceived motion states that motions as perceived and as measured are those that are not in common with the observer's motions. The observer is in a medium that determines his movements; what the observer sees as motion is what is independent of the motions he and the medium are sharing. What motions are observed, therefore, are relative to his point of view. However, what motions exist are not judged solely from this perceptual point of view, but rather from an epistemic point of view. The observer, through reason, takes himself outside the system. He cannot take God's point of view, but neither can he accept uncritically the point of view provided him by brute perception. After all it was this latter point of view that was used to support the Ptolemaic position. So the neo-Protagorean observer takes a reflective, rational point of view. The person is still the measure, but it is the rational person with critical reason who now does the judging. Starting from experience, which seemed to support the claim that the earth moves, the observer constructs a new reflexive point of view encompassing the observer and the common motions of the earth, and the body is seen to move.

Looked at in this way, Galileo's "relativity" principle is not really just a perceptual principle. It is an epistemological principle showing the conditions under which and the point of view from which humans can claim objective knowledge about motion.

Galileo pursued a strangely similar strategy in *Dialogo* where he argued that neither the Ptolemaic nor the Copernican system had a privileged point of view (perspective) with regard to the evidence of motion. No theory of natural motion (Day 1), diurnal motion (Day 2), or annual motion (Day 3) could serve as a test between the two systems. Only a motion not shared with its medium could specify a proof for one of the systems. But this motion had to be reasoned about in conjunction with the common motions it shared with other things. Galileo felt he had found the noncommon motion in the unique element water as it ebbed and flowed in the tides. Water did not share wholly in the earth's motions. Proof or objective knowledge came when the observer (Galileo) reflected upon what properties are shared with the medium and from this point of view attempts to isolate what is different. It became for the observer a process of interpretation.

This idea of interpretation by an observer and the search for standards to achieve objectivity also arose for Galileo when the "the book of nature is written in mathematics" theme was introduced in *Il Sagiatore* (1623). Mathematics (proportional geometry) would be used to interpret the book of nature, but the standard or measure by which the interpretation would be made was set by the person, from a point of view. This way of thinking also allows us to understand the primary/ secondary qualities distinction (cf. *Sagiatore*). Secondary qualities, e.g., a tickle, arose from mutual interactions between observer and environment. They were wholly perspectival and no more epistemologically significant in themselves than motions that are shared by the observer and the medium. Primary qualities, by contrast, were held to be intersubjective and knowable from many perspectives. They could be differentiated from the common and are known ultimately by reason and describable in mathematical terms (i.e., in terms of the static simple machines and equilibria models).

In an even odder way, Galileo's *Letter to the Grand Duchess* (1615) traded on this same perspectivalist, rationalized neo-Protagorean point. In that text the question was about the common or literal interpretation of the Bible. The common beliefs or perspective of biblical times had to be separated from the true (objective) intentions of scripture and interpreted according to what was known from the seventeenth-century perspective. Again the principle employed was the separation of what was common, in this case beliefs couched in the vernacular, from the objective content that was formed from Galileo's Copernican perspective.

Descartes

Numerous themes of this neo-Protagorean, perspectivalist position are adumbrated and elaborated by Descartes. Descartes' first-person philosophy of the *cogito* and the way of ideas have been much remarked in histories of philosophy. Seldom, however, has it been noted how strong a role the ego and its epistemological point of view played in Cartesian science, even in Cartesian geometry.

In a strong sense, Descartes' *Meditations* (1640) was a work that attempted to show how it was possible to base science (the objective) on first-person claims to knowledge (the subjective). Descartes argued that objectivity and knowledge of the world could be justified from the foundation of human experience, first of the *cogito* and then from reflection upon ideas. That science was possible was the conclusion of *Mediation VI*.

Another instance of the neo-Protagorean, man-is-the-measure principle is seen in Descartes' adoption of a more general version of Galileo's relativity of perceived motion principle. Descartes introduced of the relativity of motion principle in Book II of *Principia*. He contrasted movement in the ordinary sense, as an action, with movement properly speaking. Movement properly speaking was defined as

> the transference of one part of matter or of one body, from the vicinity
> of those bodies immediately contiguous to it and *considered as at rest*

(tanquam quiescentia specantur), into the vicinity of [some] others.
(Book II, Section 25; italics mine)

It was the person deciding what bodies are to be considered as at rest who determined to which bodies to attribute motion. There was no absolute motion, as Descartes further pointed out in Section 29. Translation of bodies is reciprocal, and so what counts as motion is subject to human determination. Relativity of motion and its measure based on first-person determination came together in this definition.

Later in *Principia* Book III, Descartes used the relativity principle to solve his "Copernican problem" by claiming that, strictly speaking, the earth is not moving since the surrounding matter of the vortex does not change position relative to the earth. Commonly one might talk about the earth's movement since if we consider the sun at rest, the earth is changing position relative to the sun. Like Galileo, Descartes had the observer construct a point of view outside the system in which he was living.

Another epistemological way in which this person-centered perspective came out was in Descartes' doctrine of sensation. For Descartes a person's awareness of sensations, caused by the perception of the world, did not guarantee that these sensations correspond in any way to what is really out there. Like Galileo, though in more detail, he argued that science, though ultimately relying on innate abstract common notions, must differentiate by reason what is objective from what is subjective and relative.

Even Descartes' *Geometry* (1638) was clearly person centered. Descartes' geometry differed from Galileo's. Descartes did not use the methods of proportional geometry from the static tradition. He went back to Proculus for a constructive form of geometry that required a person to rationally construct a proof. It was the person's construction of the proof that vouchsafed the proof's objectivity. In this geometry, the person as rational maker was the measure.

In a very different domain, Descartes' *Passions of the Soul* (1649) clearly exhibited his perspectivalist position. In that work Descartes set out to show the reciprocal nature of the passions. Every emotion was an action (from one point of view) and a passion (from another point of view). Emotions were the interactions of the subject (soul) and the object (body). To understand the emotions, one needed to know what caused them, and how those causes correlated with what is useful or harmful to humans. (This teleological principle relativizing everything to human good is also explicit in *Meditation VI* what it was used to state the one-to-one correspondence between bodily and mental states.) But since the emotions were both action and passion, depending upon which point of view was taken, they were one, which meant they truly were a substantial unity. This was why Descartes felt he was elucidating the unity of the mind and body in his *Passions.*

Huygens

Christiaan Huygens in his "Motion of Colliding Bodies" (*De Motu*, 1703, but probably written about 1673) represents the culmination of the Cartesian use of person-centered rhetoric in the static tradition of the mathematical problem of motion. It is interesting, since Huygens was even more aphilosophical than Galileo, that he should be the one to most clearly rely on anthropomorphic models and equilibrium principles from proportional geometry for his statical proofs about motion. This, I take it, shows how far such person-centered notions had become part of the narrative patterns of the time.

Huygens took the principle of the relativity of observed motion to be central in his presentation of the laws of collision. In his hands, however, the relativity of motion principle became a clear equilibrium principle like before in Galileo's work. Unlike Descartes or the papers on the laws of motion collected by the Royal Society (by Wren and Wallis in 1668–1669), Huygens' rhetoric clearly identified the equilibrium or balance point with a person-centered viewpoint in which the observer telling the narrative was clearly outside the system of observers watching the collision. He imagined ". . . a boat which is floating down river parallel to the bank and so close that a sailor standing in it can extend his hands to a friend standing on a bank" (*De Motu*). The sailor on the moving boat is then to move together two balls extended on strings, and his friend on the bank joining hands with him also sees the two balls coming together. However, each person sees the event of the ball's collision differently, from his own point of view.

This device of the two men and the moving boat was used throughout *De Motu*. Its language emphasized the Cartesian point that what is at rest is determined by us, is determined by a person's circumstances and point of view. It is doubtful that such a person-centered form for statical proof would have been acceptable in an earlier, non-Protagorean age.

Hobbes

Even Hobbes, the ultimate mechanist, had a person-centered epistemology deeply embedded in his philosophy. He also used the abilities and properties of the person to ground his theories about the nature of the state.

For Hobbes everything was ultimately body and motion. It would seem then that all things should be completely objective. His model of rational thought and objective knowledge was geometrical and arithmetical. But this objectivity becomes strange when we look at the nature of sensation, rationality, and value judgments in Hobbes.

Like Descartes, Hobbes argued that bodies in the world caused motions within us, and these were our perceptions. So our perception of the color of a body was nothing but motions within us that were caused by that body. Yet on Hobbes' theory, people can be objective in their color judgments. There were natural, objective standards for claims about colors even though the sensations were private and ego centered. These standards regarding the environmental circumstances for properly using color names allowed people to intersubjectively agree about color claims.

Thoughts too, for Hobbes, were nothing but motions within us. Reason was described as calculation of the sums and differences of names (and their definitions) used rightly. A person reasoned to determine how to avoid the state of nature, how to avoid evil, or (the same said differently) how to attain peace. Peace, however, was the person's desire for self-preservation. For Hobbes, what was right and wrong, good or bad was defined in terms of what was desired or to be avoided by a person. Thus the very basis for the contract, for Hobbes, was ultimately person centered (*Leviathan*, 1651, and *De Homine*, 1658). A person's sensations, emotions, reason, and actions, all, ontologically and epistemologically, lay within the individual and arose from the person's ego and its desires.

The first-person perspective, based on individual desire and natural reason, was fundamental in Hobbes, the most mechanical and objective of thinkers. This was why in his theory of human nature and individual psychology, a person's pain, pleasure, and subjective desires, were paramount and foundational.

My favorite instance of perspectivalism in Hobbes comes when he discussed love. Love was the desire for the opposite sex. But, he went on, love is called "lust" when someone disapproves (*Human Nature*). Or again, " . . . A male neighbor is usually loved one way, a female another; for in loving the former, we seek his good, in loving the latter, our own" (*De Homine*).

Spinoza

Spinoza, with his pantheism and all-encompassing, absolute God, would seem to embody the antithesis of my person-centered thesis. Yet Spinoza was the most neo-Protagorean of all, collapsing totally the concepts of idea, which was pure first-person perspective, and body, which was the object of scientific knowledge and objective. Ultimately, for Spinoza body was the object of the idea constituting the human mind (*Ethics* Book II). Affections of the body can be known only as ideas in the mind (and this was the only way mind can know body).

In other aspects of his thought, Spinoza was surprisingly like Hobbes. Not only did he take his theory of the natural motion of bodies (*Ethics* Book II) from Hobbes, but his psychology was a filtered version of Hobbes' (with a touch of Descartes thrown in). Yet there is a difference. Whereas Descartes and to a lesser degree Hobbes felt they had to describe the way in which the world acts on the human body, Spinoza explained human actions and emotions solely in terms of a person's experiences of pleasure and pain. The physiological basis and the causal connection to the world were dropped. Spinoza even derived rationality from the pleasure a person attained by desiring to think rightly. Like Hobbes too, Spinoza wrote that virtue consisted in naturally seeking one's own advantage, the highest of which was to preserve oneself (*Ethics* Book IV). Freedom was just following nature, doing what is natural.

In Spinoza again there was a person-centered epistemological basis for all knowledge of nature. True, this knowledge in some way depended upon the "axiomatic" arguments (based on Duns Scotus) that were meant to provide an understanding of God or Nature. However, the connections between Book I of the *Ethics* and what followed remained problematic and were glossed over by Spinoza and his commentators with a gratuitous appeal to a unity that is supposed to derive from his use of geometrical method. Yet from Book II onwards, the person-based focus of Spinoza's epistemology and the characteristics of individual psychology were paramount. They were the bases for arguing for the value of reason, the nature of freedom, and the structure of the social order.

Locke

Locke, the father of empiricism, should provide no epistemological surprises given his claim to base all knowledge in personal experience and his account of experience in terms of Cartesian ideas. First-person critical experience was the source of all ideas and of the patterns by which ideas were associated with one another. So it was that Locke shared the epistemological subjectivism of Descartes. Of course, there were differences, for example, the deeply probabilistic character of the nature of human knowledge as set out in the *Essay*. But this probabilism only accentuated a sceptical, yet rational, first-person perspective.

Locke's person-centered account of human understanding was intended to explain the experiential basis for the new science, which is why Locke called himself an underworker for Boyle and Newton (*Essay Concerning Human Understanding*, 1690). Locke was clearly a philosopher who believed that ego experience was the basis for scientific knowledge.

Locke's insistence on person-centered perspective was clearly shown in his critique of both the modern and the Scholastic definitions of motion. All the definitions of simple names for motion, he argued, are inadequate because they only translate the word "motion" into synonymous simple names. Real understanding of motion must come from having the simple idea of motion that comes from personal experience and then reflecting upon it.

But Locke's ultimate reliance on the first-person perspective far transcended the epistemological doctrines of the *Essay*. A very interesting extension of the neo-Protagorean perspective was his grounding of the institution of private property through the extension of his doctrine of the self and its actions. A person came to acquire property through the work or labor that made it different from land in the common state (*Second Treatise on Government*, 32). Therefore it was the person's power of constructive labor that made that land into one of his properties. Later Locke argued that government originated from a contract made by individuals owning property, all of whom had the subjective desire to preserve their property.

Locke, like Hobbes, made the person and his properties foundational for his whole theory of the state. Locke's foundation, though, was not the Hobbesian personal sensations and desires and the calculations of individual reason (though reason does play a part in the origin and sustenance of Locke's contract). The state for Locke was literally grounded upon the notion of an extension of the properties of a person by the powers of a person that included the man, his property, his family, and his slaves.

Leibniz

Perspectivalism and a peculiar version of the person-centered (or individual) thesis reached an apex in the work of Leibniz. It lay at the very basis of his visions of nature and human knowledge and in his emphasis on the relative and relational character of science. Leibniz was in many ways the ontological and epistemological culmination of the seventeenth-century neo-Protagorean tradition, yet he strove mightily to bring necessity and objectivity into his world.

Ontologically Leibniz's perspectivalist position was best seen in his later metaphysics of monads. The monads, like Spinoza's God, were pure activity. Each was a pure ontological subject, or individual, unaffected by other monads and wholly tied into its own point of view. Passivity for Leibniz was a defined state that came about by a monad's reflecting on the point of view that other monads took toward itself. The key concepts for describing these basic monads were force (*vis*), which defined their activities, and desire or tendency (*conatus*), which described the laws by which monads operated.

At the metaphysical level, monads and their powers and tendencies grounded the activities and powers of things existing in the phenomenal world of bodies and people. The set of infinite relations between the phenomenal level and the metaphysical level of monads meant that human knowledge could not completely grasp the grounds or the connections between these levels. Human knowledge was always incomplete and always probabilistic (*Monadology*, 1714).

At the phenomenal level, a body in motion was characterized as having a living force that caused motion and acceleration (*vis viva*). This force was ultimately a manifestation of forces that existed at the metaphysical monadic levels. It was only this latter force that was ontologically real (*Specimen Dynamicum*, 1695). Motion, as humans saw it in the mundane bodies around them, occurred in space and time, but these too were only relational epistemological concepts that were defined for our own human good in order to allow for limited human comprehension of and communication about motion. Time was not real but only a measure by which humans thought about motion. Time for Leibniz, as for Galileo, Descartes, and even Aristotle, was literally the measure of motion, a person-based measure. Time was a human, phenomenal level, concept-grounded in the exfoliation or succession of monadic states. Space was the measure humans use to perceive relations among bodies. These relations were necessary, so that harmony and noncontradiction could hold among the careers and points of view of the different monads and, as far as humans could know, among the laws of nature.

The upshot of these metaphysical and phenomenal doctrines was a conception of the relativity of motion that was complete and shared none of the vagaries or problems of Descartes. All hypotheses about which bodies were in motion and which at rest had to be equivalent, and the equivalency had to be reflected in the laws of collision. Not even an angel could tell which body was really in motion (*Specimen Dynamicum*).

So Leibniz's version of the mechanical philosophy as the way to look at science, along with its basis in his unique metaphysics, tied together the perspectivalism of the first person with the relativity of motion, the two tendencies that we saw nascently in Galileo and then side by side, though with tension, in Descartes.

Newton

Newton, like Leibniz, had an ontology of forces, and like Spinoza, he was a theist. But the grounding of the forces for Newton was not in metaphysical monads nor in the identification of a rational God with nature, but directly in an all-powerful, voluntaristic God. His belief in this simpler metaphysics allowed Newton to think that he could bring objectivity back into science. In fact in the *Principia* (1687), Newton felt he was quite entitled to suppress any of the fundamental ideas that played a part in his understanding of the nature of force or that were presupposed by his concept of motion.

Newton argued that space was an emmanent effect or attribute of God. Space was structured by an infinite number of possible shapes, each overlapping with others. God, in creating the world, chose freely which of these shapes in space he would activate or fill with force. The filling of a selection of shapes in space constituted the creation of matter, and the force that God put into these shapes was the force of resistance or impenetrability (*De Gravitatione et Aequipondio Fluidorum*, 1664–1668). This force from God became the *vis insita* of the *Principia*. Space then, like God, was something real, something absolute.

Given the absolute nature of space and the reality of forces, there was no room in Newton's system for relativistic motion or for any person-centered epistemological worries. In Newton's system the measure of all things was removed from human beings and placed totally in God and his will. But paradoxically, this voluntaristic God acted by his own necessity and grounded a world and knowledge that were completely absolute and objective. Given real forces and absolute space and time, humans could make apodictic claims about the nature of motion and about the structure of the world. Once this grounding was understood (as it was by Newton), the writer of science never needed again to make reference to the epistemological problems that led Descartes and his followers into their person-centered concerns. In fact, as Newton argued in *De Gravitatione*, it was the doctrine of the relativity of motion that caused Descartes his troubles. So if one abandoned relativity, by the ploy of the all-powerful will of God, one abandoned thereby Descartes' errors. Objectivity had returned to science. Humans could know everything, including the absolute character of motion (which was the point of the rotating bucket experiment in *Principia*).

Conclusion

In the *Principia* Newton completely divorced the presentation of his science from any concern about the nature of humans or their knowledge. For Newton man was not the measure of anything. Everything was measured in absolute terms, ultimately by God. At the end of the seventeenth century, there were certainly other concurrent reasons and conditions that abetted this divorce of science from the rest of the intellectual world, including theology, metaphysics, epistemology, psychology, and the like. But whatever the manifold causes, the person-centered, neo-Protagorean character of science was abandoned. There was no need to take into account the nature of the human being and reason when discussing the nature of the world. Newton had made the world into an objective place.

The Rhetoric of Certainty: Newton's Method in Science and in the Interpretation of the Apocalypse

MAURIZIO MAMIANI

I. The Rhetoric of Certainty

Cartesian inquiry on method has greatly emphasized the problem of certainty of knowledge. It is known that mechanical hypotheses, according to Descartes, do not disprove certainty: they simply do not count on it. Certainty pertains to the *congruence* of hypotheses and experience, namely, to the consequences that come from the explicative hypotheses.[1]

Thus the hypotheses' uncertainty in no way affects physical knowledge. It is possible to explain reality the way it seems easier (the clearer and more distinct way for the thought) without fearing the falsity or uncertainty of the hypotheses that have led us to that explanation. One passage of the *Dioptrique* explains efficaciously and clearly this fundamental assumption of Cartesian methodology:

> Now, having here no other occasion to speak about light, except that to explain how its rays enter the eye, and how they may be deflected by different bodies they meet with, I need not undertake to say what is really its nature, and I think it will be sufficient I shall use two or three comparisons which may help to conceive it the way it seems to me

easier to explain all of its properties that experience make us to know, and afterwards to deduce all others that may not be observed so easily; thus imitating astronomers who, though almost all their hypotheses were false or uncertain, nevertheless, since they apply to various observations made by them, do not fail to draw several most true and certain consequences.[2]

Newton did not accept this Cartesian assumption, protesting on many occasions the crucial linking between hypotheses and uncertainty in knowledge.

On 11 June 1672, Newton, replying to some of Hooke's objections to his new theory of colors, asserts that he chose to decline all mechanical hypotheses, since it was impossible to explain physical properties by one hypothesis alone.[3] Another letter, written most likely on the same day to Pardies, explains the reason for Newton's choice:

And if anyone makes a guess at the truth of things by starting from the mere possibility of hypotheses, I don't see how to determine any certainty in any science; if indeed it be permissible to think up more and more hypotheses, which will be seen to raise new difficulties.[4]

Certainty is more important than explanation. What is the use of putting forward an ingenious explanation on the uncertain ground of any hypothesis whatever?

Since then, the refusal of hypotheses became a basic and recurring theme for the constitution of Newton's scientific method, until the well-known *Hypotheses non fingo* in the *Scholium generale* (1713) of the *Principia*. No doubt it was a negative theme, but it had strong effects: any knowledge has a certainty level reciprocally proportional to the hypothetical status of its premises or principles. As Newton points out in the same letter to Hooke, " . . . the absolute certainty of a Science cannot exceed the certainty of its Principles."[5]

Newton overturned the basic methodological assumption of Descartes, who believed that the certainty of a science is independent of the certainty of its principles, the clearness and distinction of explanations being sufficient.

Newton's science was thus against rhetoric: it did not need to seek assent or persuasion for its own conclusions. Once hypotheses are rejected, the level of opinion is given up, and uncertainty is eluded. Experiments "concluding directly & wthout any suspicion of doubt"[6] and mathematical abstraction take the place of opinions.

In one passage of the *New Theory about Light and Colors*, omitted from the printed *Philosophical Transactions*, Newton's aim of opposing rhetorical argumentations is explicit:

For what I shall tell concerning them is not an Hypothesis but most rigid consequence, not conjectured by barely inferring 'tis thus because not otherwise or because it satisfies all phaenomena (the Philosophers universall Topick).[7]

Therefore, Newton argues he has good reasons for asking Oldenburg to help prevent the opponents of his theory from involving him in verbal disputes and even from raising any objection (the objection was a canonical figure of the Aristotelian [8] rhetoric and topic):

And therefore I could wish all objections were suspended, taken from Hypotheses or any other Heads than these two; Of showing the insufficiency of experiments [. . .]; Or of producing other Experiments wch directly contradict me, if any such may seem to occur.[9]

Thus experiment performs the functions that traditionally were assigned to rhetoric: whereas the orator manifested by enthymemes the insufficiency of an opponent's arguments or advanced objections against his premises, so now Newton requires that discussion on scientific conclusions be performed, positively or negatively, by experiments.

Newton's confidence is based on his implicit conviction that experiments, unlike hypotheses, have so high a degree of unambiguousness as to avoid uncertainty.

However, the various objections to his theory of colors seemed to testify to the contrary. Hooke, for example, was willing to declare that Newton's experiments were actually correct, but he also asserted that Newton's theory did not follow directly and without any suspicion of doubt from them. To Hooke it seems "more naturall"[10] (Descartes would say "easier," i.e., clearer and more distinct) to think that colors' separation results from a mechanical modification caused by the refracting medium, rather than to ascribe the phenomenon to original and immutable qualities of light rays.

Hooke's criterion, namely, to choose the "more naturall" way of thinking, leads to the question: is it possible to decide what is the "more naturall" way of thinking?

Facing Hooke's objections and then Huygens',[11] Newton appeals to factual evidence, which he is convinced is clearly separable from any hypothesis:

I never intended to show wherein consists the nature and difference of colours, but onely to show that *de facto* they are originall & immutable qualities of the rays wch exhibit them, & to leave it to others to explicate by Mechanicall Hypotheses the nature & difference of those qualities.[12]

But it is precisely about this that both Hooke and Huygens disagree with Newton: the attribute of original and immutable quality could not be a fact, only a hypothesis. Surely it was not a *mechanical* hypothesis, but neither was it a fact. They correctly saw this opinion as the (hypothetical) explanation of a fact, also, and chiefly, since it did not pretend to disclose the nature and difference of colors.

Koyré[13] has ironically drawn our attention to the fact that the word "hypothesis" becomes in Newtonian science a singular word, like the word "heresy": only other men, not Newton, feign hypotheses; only other men, who do not profess our own faith, are "heretical."

Newton's next effort is sketched in the same letter to Huygens; it is the attempt to axiomatize his theory of colors in a mathematical way.[14]

On 21 September 1672, writing to Oldenburg, Newton had already anticipated his intention of using this method:

> To comply with your intimation . . . I drew up a series of such Expts on designe to reduce ye Theory of colours to Propositions & prove each Propositions from one or more of those Expts by the assistance of common notions set down in the form of Definitions & Axioms in imitation of the Method by wch Mathematitians are wont to prove their doctrines.[15]

Thus Newton prefers to imitate the method of mathematicians rather than, as Descartes had put forward, that of astronomers!

According to the explicit expression of Newton himself, this method was an attempt to avoid "some misunderstanding."[16]

I argue on the contrary that it was a rhetorical device: Newton utilized an enthymeme directly against Huygens' hypothesis that two colors were sufficient to compound all others, and indirectly against mechanical explanations of Hooke. Newton's enthymeme is easily recognizable in the following four propositions in his letter to Huygens:

> 1. The Sun's light consists of rays differing by indefinite degrees of refrangibility. 2. Rays wch differ in refrangibility, when parted from one another do proportionally differ in the colours wch they exhibit. These two Propositions are matter of fact. 3. There are as many simple or homogeneal colours as degrees of refrangibility. . . . 4. Whiteness in all respects like that of the Sun's immediate light & of all ye usuall objects of our senses cannot be compounded of two simple colours alone. For such a composition must be made by rays that have onely two degrees of refrangibility by Def. 1 & 3; & therefore it cannot be

like that of the suns light by Prop. 1; Nor for ye same reason like that
of ordinary white objects.[17]

It is possible to dispose of these four propositions according to the classical
structure of a deductive syllogism, except that Newton regards the first two
propositions as facts. At any rate, Newton is convinced that he has been using a
mathematical method, not a dialectical or rhetorical one.

Newton will call arguments, such as this one, reasons or proofs, but their
premises are always derived *de facto* from experiments. In this manner Newton
believed that he had banished the uncertainty of hypotheses, giving at last an
adequate method to natural science.

However, the proofs used by Newton are only formally mathematical, since
the premises that he assumes are not endowed with logical necessity.

Newton believes that the certainty of his premises is evinced from the
certainty of experiments, but Hooke and Huygens could object that the latter does
not involve the former because premises are not evinced directly from facts, and
can never be evinced from them because premises are hypothetical explanations.
Thus the premises of Newton's enthymeme would also be simply made of
opinions: the proof of composition of white light would result in a confutation of
an opponent's opinion, drawn from other plausible premises.

Consequently Newton must admit that his conjecture is based on the inference that it is thus because it is not otherwise. Huygens' argument is not directly
confuted by experimental evidence (the matter of fact), but by theoretical hypotheses that are implied in the initial propositions. The strength of his argument
is wholly founded on the rhetorical device of confutation: Newton's proofs or
reasons are equivalent to the enthymemes that Aristotle delineates in his *Rhetoric*.
The enthymeme, namely, the rhetorical syllogism, is the demonstration proceeding from plausible premises. According to whether these premises are or are not
accepted by the interlocutor, the enthymeme is either demonstrative or confutative.
Technically it is identical to the apodictic syllogism, from which it differs because
its premises are not necessary.[18]

The Newtonian problem of isolating uncertainty becomes another problem,
namely, that of establishing what is the most natural way of thinking (of formulating the initial hypotheses); but it is impossible to resolve this problem if one, as
Newton does, refuses to discuss the theoretical premises. What criterion does one
use to choose between two premises that will ensure that one can support the true
one or at least the best one?

Newton furnished the ultimate reply to this query in the last *regula
philosophandi* of the third edition of the *Principia*. But it was an unsatisfying reply
because it made a rule of the rhetorical device, which stated that the premises

evinced from facts are to be regarded as facts, and that it is impossible to derive other premises from them.

II. The Interpretation of Prophesies

The congruence between Newton's scientific convictions and his religious beliefs may appear fortuitous. And yet the methodological affinity between scientific and religious knowledge is so narrow in Newton's works that it seems highly improbable that chance is at work. Both science and religion have a common root and a common reference to certainty, to which they come by means of devices that can eliminate hypotheses: as we have seen, this is precisely what characterizes Newton's methodological approach. Moreover, a comparison between Newtonian scientific method and his way of interpreting prophesies further emphasizes the relevant task that rhetoric plays in both. Here I shall sketch this comparison by examining Newton's vast treatise, handwritten, partially transcribed by Frank E. Manuel,[19] about the interpretation of the *Apocalypse*.

The interpretation of the prophesies is more important than that of nature; sure enough, Newton says, "this is no idle speculation, no matter of indifferency but a duty of the greatest moment."[20]

Both the individual's destiny and the church's salvation need a certain interpretation of the prophesies:

> That the benefit wch may accrew by <ye> understanding the sacred Prophesies & the danger by neglecting them is very great & that ye obligation to study them is as great may appear by considering ye like case of ye Jews at ye coming of Christ. For the rules whereby they were to know their Messiah were the prophesies of the old Testament.[21]

Newton carries on his argumentation by a parallelism: as the Jews had to distinguish Christ by prophesies, so also will the Christians have to recognize Antichrist. And as the first were punished for not giving importance to the prophesies of Christ's first coming, the second will not be excused for neglecting those of the last ages:

> And If God was so angry with ye Jews for not searching more diligently into ye Prophesies wch he had given them to know Christ by: why should we think he will excuse us for not searching into ye Prophesies wch he hath given us to know Antichrist by? For certainly it must be as dangerous & as easy an error for Christians to adhere to Antichrist as it was for ye Jews to reject Christ.[22]

Whence comes the necessity that the interpretation of the prophesies be sure. But how can we avoid the uncertainty of interpretations? Newton's strategy is the same he followed in the defense of his scientific method. Having underlined the relevance of the matter, Newton sketches the appropriate method for interpreting the prophesies:

> Considering therefore the great concernment of these scriptures & danger of erring in their interpretation, it concerns us to proceed wth all circumspection. And for that end I shall [propound to my self] make use of this Method. First I shall lay down certain [Rules] general Rules of Interpretation, ye consideration of wch may prepare the judgment of ye Reader & inable him to know when an interpretation is genuine & of two interpretations which is the best.[23]

Newton's primary aim is to elude uncertainty:

> By wch means the Language of ye Prophets will [appear] become certain & ye liberty of wresting it to private imaginations be cut of. The heads to wch I reduce these words I call Definitions.[24]

Private imaginations are manifestly equivalent to hypotheses. Being thus designated the Rules and Definitions, the mechanism of demonstration is now ready:

> These things being premised, I compare ye pts of ye Apocalyps one wth another & digest them into order by those internal characters wch ye Holy-ghost hath for this end imprest upon them. And this I do by drawing [them] up the substance of ye Prophesy into Propositions, & subjoyning the reason for ye truth of every Proposition.[25]

The demonstration of these propositions is obtained through proofs or reasons that definitions and rules make possible: it is the same device that in 1672 Newton called the mathematical way and proposed to Huygens as a way of avoiding "some misunderstanding" about Newton's theory of colors. This device, as I have said, does not produce by itself mathematical demonstrations, but enthymematic conclusions in which demonstrative form has an intrinsically rhetorical function. The dissimilarity among the objects to be investigated (be they prophesies or physical causes) induces no sensible alteration of the method proposed by Newton to obtain certainty. Thus the rules for interpreting the words and the language of Scripture are reciprocal with the *Regulae philosophandi* of the *Principia*.

The accord of Scripture and the analogy of prophetic style are equivalent to the simplicity of nature; and as one must assign the same causes to natural effects of the same kind, so one must ascribe one meaning, and one only, to a single passage of Scripture, maintaining as far as possible the same meaning of the words.

As in the third rule of the *Principia*, an attested meaning has to be extended to all of Scripture:

> To acquiesce in that sense of any portion of Scripture as the true one wch results most freely & naturally from ye use & propriety of ye Language & tenor of the context in that & all other places of Scripture to that sense. For if this be not the true sense, then is the true sense uncertain, & no man can attain to any certainty in ye knowledg of it. Which is to make ye scriptures no certain rule of faith, & so to reflect upon the spirit of God who dictated it.[26]

Uncertainty becomes in this case the very source of heresy because private opinion is here preferred to the divine authority of Scripture:

> He that without better grounds then his private opinion or the opinion of any human authority whatsoever shall turn scripture from the plain meaning to an Allegory or to any other less naturall sense declares thereby that he reposes more trust in his own imaginations or in that human authority then in the Scripture [& by consequence that he is no true beleever]. And therefore ye opinion of such men how numerous soever they be, is not to be regarded. hence it is & not from any reall uncertainty in ye Scripture yt Commentators have so distorted it; And this hath been the ye door through wch all Heresies have crept in & turned out ye ancient faith.[27]

Also the fourth rule of the *Principia* may be recalled: as a proposition derived by induction must be assumed to be true, unless other phenomena do not arise to make it more exact, so in the interpretation of Scripture "if two meanings seem equally probable he is obliged to beleive no more then in general ye one of them is genuine untill he meet wth some motive to prefer one side.[28]

Manuel dates these first commentaries of the prophesies to the 1670s and 1680s,[29] when Newton had not yet written the *Regulae philosophandi*. In this case chronology is scarcely relevant. The affinity in the Newtonian quoted passages is sufficient to suggest that Newton has been engaged in the same problem (to avoid uncertainty) and that he has been using the same method (to banish hypotheses and private imaginings) both in interpreting Scripture and in practicing science. Newton moves forward according to the presumption that methodological rules reveal the simplicity both of nature and Scripture.

This common assumption is explicitly declared in another rule for interpreting the *Apocalypse*:

To [prefer] choose those [interpretations] constructions wch without straining reduce things to the greatest simplicity. The reason of this is manifest by the precedent Rule. Truth is ever to be found in simplicity, & not in ye multiplicity & confusion of things. As ye world, wch to ye naked eye exhibits the greatest variety of objects, appears very simple in its internall constitution when surveyed by a philosophic understanding, & so much ye simpler by how much the better it is understood, so it is in these visions. It is ye perfection of [all] God's works that they are all done wth ye greatest simplicity. He is ye God of order & not of confusion. And therefore as they that would understand ye frame of ye world must indeavour to reduce their knowledg to all possible simplicity, so it must be in seeking to understand these visions. And they that shall do otherwise do not onely make sure never to understand them, but derogate from ye perfection of ye prophesy; & [declare] make it suspicious also that their designe is not to understand it but to shuffle it of & confound ye understandings of men by making it intricate & confused.[30]

Having stated these rules, Newton argues that he has not only defined the appropriate method for interpreting the words and language in Scripture, but has also actually interpreted them according to this method. Perhaps remembering what happened with his theory of colors, Newton anticipates his opponents with almost the same words by which he defended himself from Hooke's objections, namely, by refusing to consider alternative explanations:

Hence if any man shall contend that my Construction of ye Apocalyps is uncertain, upon pretence that it may be possible to find out other ways, he is not to be regarded unless he shall show wherein what I have done may be mended.[31]

Newton moves forward using Hooke's expression "naturall" to indicate the stronger reason for establishing truth:

If ye ways wch he contends for be *less natural* or grounded upon weaker reasons, that very thing is demonstration enough that they are fals, & that he seeks not [after] truth but [labours for] ye interest of a party.[32]

Once more, what is the most natural way of thinking is the question. And Newton clearly reveals the rhetorical aim of his methodology, since no other demonstration is needed to persuade the reader:

And if ye way wch I have followed be according to ye nature & genius of ye Prophecy there needs no other demonstration to convince it [the reader].[33]

Thus the method is sufficient to guarantee the certainty of what is known by it, and the possible objections pertain only to its application (to show the insufficiency of experiments or to correct the interpretations that do not follow the rules). Therefore, Newton has the conviction that his interpretation of *Revelations* cannot "but move ye assent of any humble & indifferent person that shall wth sufficient attention peruse them & cordially beleives the scriptures."[34]

An analogous statement can be made about the scientific method that he was defending from the objections of Hooke and of Cartesian followers. It is strange that this defense is conducted by the same argument used by Hooke: the method proposed by Newton is the more *natural* way of reasoning or, as it is said in the *Opticks*, the best way of reasoning that the nature of things allows.[35]

But it is clear that Newton plays on words, since the nature of things is ascertained by way of reasoning: the premises derived by induction are hypothetical if referred to a natural way of reasoning or to the nature of things. There is no warranty that the same way of reasoning or the nature of things does not lead us to hypotheses that can explain all the known phenomena.

Thus the problem of the certainty of hypotheses is shifted by Newton to that of the rightness of conclusions, which may be proved by formal demonstrations.

It is significant that Newton makes a comparison between his construction of the *Apocalypse* and the assembly of an engine:

> For as of an Engin made by an excellent Artificer [every] a man readily beleives yt ye parts are right set together when he sees them joyn truly with one another notwithstanding that they may be strained into another posture; & as [every] a man [readily] acquiesces in ye meaning of an Author how intricate so ever when he sees ye words construed or set in [the] order according to ye laws of Grammar, notwithstanding yt [ye words may possibly be forceing] there may be a possibility of forceing ye words to some other harsher construction: so a man ought wth equal [construction] reason to acquiesce in [that] the construction of these Prophesies when he sees their parts set in order according to their suitableness & the characters imprinted in them for that purpose. Tis true that an Artificer may make an Engin capable of being wth equal congruity set together more ways then one, & that a sentence may be ambiguous: but this Objection can have no place in the Apocalyps, becaus God who knew how to frame it without ambiguity intended it for a rule of faith.[36]

In this case, Newton, who knows he is moving on the uncertain grounds of belief, wants to persuade the reader that it is always possible to specify an interpretative structure analogous to an engine in which all the parts obey exact laws of construction, just as an author's words are ordered according to the laws of grammar. It is the same for physical phenomena, which obey the "nature" of things.

Actually Newton is proposing in his turn a giant rhetorical engine, in which the possibility, recognized only by words, of a different construction of parts endowed with equal congruence is annihilated by the assumption that God may not be ambiguous or that nature may not be equivocal. No chance is given to objections, which in such a case may only derive from ignorance or partiality. The other face of the Newtonian search of certainty is that of forbidding objections. We cannot deny that Newton's arguments were persuasive, since they have convinced most of the scientific community for almost 300 years!

But they are rhetorical arguments: resorting to God as a warrant for the congruence of prophesies or appealing to the nature of things as the foundation of the unambiguity of facts is the principal rhetorical device by which Newton wants to capture the reader's assent. The order of the whole is the justification of the truth of the parts, but parts have to be chosen before order is found. Demonstration leads us to certainty, and it is sufficient to prove the truth of premises.

III. Newton's Scientific Method as a Rhetorical Device

It is quite legitimate for Newton to try to defend his scientific conclusions or the results of his interpretation of *Revelations* by means of rhetorical arguments. But Newton wishes to make his interpretation the final one, without allowing others what he has claimed for himself: the liberty to choose new theoretical hypotheses.

Thus Newton conceals the intense activity that prepares scientific discovery, which is made of uncertain hypotheses and approximate trials: the roots of invention and of innovation are cut off, avoiding the uncertainty of knowledge.

As Sabra has pointed out,[37] the *experimentum crucis* of the *New Theory* produces an almost hypnotic effect. It is convincing on the basis of a rhetorical appeal to facts, which conveys the belief that the gap between facts and theory does not exist.

In addition to this hypnotic appeal to facts, geometrical demonstrations, according to the ancients' style, are the second rhetorical device of Newtonian science. In Aristotle's *Rhetoric*, the demonstration is already suggested as a highly persuasive issue: "In fact we give credit particularly to what we presume to have been demonstrated."[38]

It is quite apparent that Newton is proposing a scientific method whose principal task is supposed to capture factual and demonstrative certainty. Thus Newton did not limit himself to the use of various rhetorical devices in order to defend his own discoveries. By means of the geometrical demonstration (propositions, lemmas) proceeding from more or less factual premises (definitions drawn from phenomena and linked to propositions by general rules), he built up a rhetorical device so efficient that it needed no other persuasive argument.

As Newton had written in the passage of the *New Theory* omitted from the printed version, and as he reminded Hooke,[39] the science of colors was mathematical and as certain as any other part of optics. Newton had made an admirable methodological synthesis of the characteristics the new science had expressed up to that time: the experimental approach and the philosophical use of mathematics. Nevertheless, his method was clearly linked with more ancient cultural traditions, logic and rhetoric included.

In one passage of MS Add. 3968, perhaps belonging to a projected preface to the *Principia*, Newton quotes the ancients and their mathematical methods:

> The Ancients had two Methods in Mathematicks wch they called Synthesis & Analysis, or Composition & Resolution. By the method of Analysis they found their inventions & by the method of Synthesis they [published them] composed them for the publick. The Mathematicians of the last age have very much improved [Analysis &] Analysis [& laid aside the Method of synthesis] but stop there [in so much as] & think they have solved a problem when they have only resolved it, & by this means the method of Synthesis is almost laid aside. The Propositions in the following book were invented by Analysis. But considering that [they were] the Ancients (so far as I can find) admitted nothing into Geometry [but wha] before it was demonstrated by Composition I composed what I invented by Analysis to make it [more] Geometrically authentic & fit for the publick.[40]

The equivalence between "geometrically authentic" and "fit for the publick" may astonish the modern reader: actually it denotes the rhetorical meaning that demonstration already had according to Aristotle.

Therefore, in interpreting *Revelations*, Newton also utilized the geometrical method in its rhetorical function to "prepare the judgment of ye Reader & inable him to know when an interpretation is genuine & of two interpretations which is the best."[41]

The distinction between analysis and synthesis made by Newton corresponds well enough to the modern one between discovery and justification, but it also recalls analogous distinctions in the traditional logical and rhetorical treatises.

Newton certainly was well acquainted with Robert Sanderson's *Logicae artis compendium*, since he had given it careful study as young Cambridge undergraduate. Sanderson makes a distinction between the method of discovering knowledge and the method of presenting or teaching it. The first of these is called the method of invention and the second the method of doctrine, which may be divided into two procedures: the "resolutiva" and the "compositiva."[42]

These two procedures were applied to the theoretical sciences (the "compositiva") and to the practical ones (the "resolutiva"). Sanderson lists five laws or rules as common to both, and two laws as peculiar to each. A comparison between these laws and the *regulae philosophandi* or the rules for interpreting the *Apocalypse* leaves no doubt that Newton took hints from Sanderson's *Compendium*.

Among the five laws common to the two methods, the first is called *Lex brevitatis* (nothing should be superfluous in a discipline) and the second *Lex harmoniae* (the parts of a doctrine should agree among themselves): both laws have been used by Newton. The first two *regulae philosophandi* are a good arrangement of the *Lex brevitatis* applied to phenomena; the second *regula philosophandi* also exhibits some affinity with the third law listed by Sanderson, namely, the *Lex unitatis, sive homogeniae* (no doctrine should be taught that is not homogeneous). The foundation of the rules for interpreting the *Apocalypse* is, as we have seen, the *Lex harmoniae* of Sanderson.

Sanderson's *Compendium* combines the scholastic matter of the logical and rhetorical tradition corresponding to the liberal arts of the trivium. Rhetoric, according to Sanderson, directs logic and grammar, being almost the synthesis of both, and is charged with moving the passions. Taken together, they represent the discursive (grammar and, in part, rhetoric) and the rational (logic and, in part, rhetoric) side of philosophy.[43]

Newton forgets all these scholastic distinctions, except that between *resolutio* and *compositio*, and considers the latter fit for the public. In the *compositio*, the demonstrative form prevails in the examination of premises, and their potential uncertainty is concealed.

Persuasion should necessarily follow the application of this method: the less the utilization of hypotheses, the larger the certainty of conclusions, and hence the larger their hypnotic effect. If hypotheses were wholly rejected, certainty would reach its climax, and there would be no want of rhetorical devices for obtaining persuasion. Every indifferent man would give consent to the evidence of facts or to the rightness of demonstrations.

This is the method followed by Newton in the *Opticks*, in the *Principia*, and in the interpretation of the *Apocalypse*. Thanks to it, theoretical discussions were made useless, objections not conforming to the same method unacceptable, and private imaginings definitively annihilated.

Modern science has transformed into a paradigm this Newtonian knowledge ideal, fancying that the objectivity of knowledge may be warranted from the correct employment of a method, in which reason and experience seem to reflect

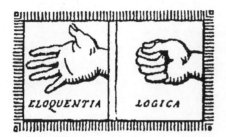

Rhetoric and logic. (From
Wilbur Samuel Howell,
*Logic and Rhetoric in
England, 1500–1700*
(New York: Russell &
Russell, 1961)

each other. And yet Newton to build this science had sacrificed dialectics, the art of disputing, to the method, assigning to the latter the cogency and lucidity of the former; meanwhile, the method was propounded as a form of expression more convincing for the public and reader, that is, in the rhetorical meaning according to Sanderson, Ramus, and the Scholastics.

This exchange of characteristics between dialectics and rhetoric was not without consequences for the advancement of science. In scholastic treatises, rhetoric was sometimes pictured as an open hand, and logic or dialectic as a fist.

Surely Newton left the legacy of science closed in its demonstrative processes because of the fear of uncertainty in its hypothetical grounds and verbal disputes that the open hand seemed to indicate. But the open hand also represented the inventive might of the discussion and of the expression, which was exercised by Newton too. However, these activities were extraneous to the fanatical insistence on certainty claimed by Newton himself.

The oscillation of scientific method between logic and rhetoric, between open and clenched hand, may be understood as the consequence of the difficult convergence of two ancient cultural traditions, which the new science upset by significantly modifying the relations among disciplines.

Surely modern science has never recognized its debt to rhetoric, claiming to be not in need of it, just when scientific method makes use of it in a hyperbolic manner.

NOTES

1. *Oeuvres de Descartes*, 12 vols., Charles Adam and Paul Tannery, eds. (Paris: Vrin, 1964), VI, p. 76; VIII–1, pp. 100–101.

2. "Or, n'ayant icy autre occasion de parler de la lumiere, que pour expliquer comment ses rayons entrent dans l'oeil, & comment ils peuvent estre détournés par les divers

cors qu'ils rencontrent, il n'est pas besoin que i'entreprene de dire au vray quelle est sa nature, & ie croy qu'il suffira que ie me serve de deus ou trois comparaisons, qui aydent a la concevoir en la façon qui me semble la plus commode, pour expliquer toutes celles de ses proprietés que l'experience nous fait connoistre, & pour deduire en suite toutes les autres qui ne peuvent pas si aysemente estre remarquées; imitant en cecy les Astronomes, qui, bien que leurs suppositions soyent presque toutes fausses ou incertaines, toutefois, a cause qu'elles se rapportent a diverses observations qu'ils ont faites, ne laissent pas d'en tirer plusieurs consequences tres vrayes & tres assurées," René Descartes, *La Dioptrique*, in *Oeuvres*, VI, p. 83. See also, Maurizio Mamiani, *Il prisma di Newton* (Roma-Bari: Laterza, 1986), pp. 5–9.

3. "But I knew that the Properties wch I declared of light were in some measure capable of being explicated not onely by that, but by many other Mechanicall Hypotheses. And therefore I chose to decline them all . . . ," *The Correspondence of Isaac Newton*, H.W. Turnbull, ed. (Cambridge: Cambridge University Press, 1959), I, p. 174.

4. "Et si quis ex sola Hypothesium possibilitate de veritate rerum conjecturam faciat, non video quo pacto quicquam certi in ulla scientia determinare possit; siquidem alias atque alias Hypotheses semper liceat excogitare, quae novas difficultates suppeditare videbuntur," *ibid.*, p. 164.

5. *Ibid.*, p. 187.

6. *Ibid.*, p. 97.

7. *Ibid.*

8. Cf. Aristotle, *Rhetorica*, 1402b1.

9. Newton, *The Correspondence*, I, p. 210.

10. *Ibid.*, p. 111.

11. "Je ne vois pas aussi, pourquoy Mr Newton ne se contente pas des 2 couleurs, Jaune et Bleu. Car il sera bien plus aisé de trouver une hypothese par le mouvemt, qui expliquat ces deux differences, que non pas pour tant de diversitez qu'il y a d'autres couleurs," *ibid.*, p. 255.

12. *Ibid.*, p. 264.

13. A. Koyré, "L'Hypothèse et l'experiénce chez Newton," *Bullettin de la Societé Française de Philosophie* 50 (1956), p. 79.

14. Newton, *The Correspondence*, I, p. 292–293.

15. *Ibid.*, p. 237. Cf. M. Mamiani, *Isaac Newton filosofo della natura* (Firenze: La Nuova Italia, 1976), pp. 184–212.

16. Newton, *The Correspondence*, I, p. 292.

17. *Ibid.*, p. 293.

18. "There are two kinds of enthymemes: indeed some prove what is or is not, others are confutative, and differ as in dialectic syllogisms the confutation differs from syllogism. In fact demonstrative enthymeme is that which proceeds from accepted

premises; and the confutative one draw conclusions from others which are not accepted," Aristotle, *Rhetorica*, 1396b22–28. Cf. also 1356b and 1357b6.

19. Frank E. Manuel, *The Religion of Isaac Newton* (Oxford: Clarendon Press, 1974). I have transcribed from the original ms. the passages quoted afterwards.

20. Jewish National & University Library, Jerusalem, Yahuda MS 1, f. 3r.

21. *Ibid.*, f. 2r.

22. *Ibid.*, f. 3r.

23. *Ibid.*, f. 9r–10r.

24. *Ibid.*, f. 10r.

25. *Ibid.*

26. *Ibid.*, f. 12r–13r.

27. *Ibid.*, f. 13r.

28. *Ibid.*, f. 12r.

29. Frank E. Manuel, *The Religion of Isaac Newton*, p. 14.

30. Yahuda MS 1, f. 14r.

31. *Ibid.*

32. *Ibid.*, f. 14r–15r. Italics added.

33. *Ibid.*

34. *Ibid.*

35. Isaac Newton, *Optics* (Chicago: University of Chicago Press, 1952), p. 543.

36. Yahuda MS 1, f. 15r.

37. A. I. Sabra, *Theories of light from Descartes to Newton* (London: Oldbourne, 1967), p. 249.

38. Aristotle, *Rhetorica*, 1355a.

39. Newton, *The Correspondence*, I, p. 187.

40. University Library, Cambridge, MS Add. 3968, f. 101.

41. Yahuda MS 1, f. 9r.

42. R. Sanderson, *Logicae artis compendium* (Oxoniae, 1618), pp. 226–227. In Newton's library, there was the third edition of Sanderson's *Compendium* (Oxoniae, 1631) with Newton's signature and date ("Isaac Newton Trin Coll Cant 1661") and a few signs of dog-earing. Cf. John Harrison, *The Library of Isaac Newton* (Cambridge: Cambridge University Press, 1978), p. 231.

43. Cf. Wilbur Samuel Howell, *Logic and Rhetoric in England, 1500–1700* (New York: Russell & Russell, 1961), p. 304.

Quanta, Relativity, and Rhetoric*

GERALD HOLTON

I. By Way of Prologue

Rhetoric in Science? To a scientist, the very phrase has all the signs of an oxymoron. Since ancient times rhetoric has been essentially the art of persuasion, in contrast to the art of demonstration. Of all the claims of modern science, perhaps the strongest is to have achieved, in painful struggle over the past four centuries, an "objective" method of demonstrating the way nature works, of finding and reporting facts that can be believed regardless of the individual, personal characteristics of those who propose them, or of the audience to which they are addressed. This distinction of the roles of objectivity and subjectivity is clear in Aristotle's *Rhetorica*:[1] Of the three kinds of "modes of persuasion" available to the speaker relying on rhetoric, only the third "depends on the proof, or apparent proof, provided by the words of the speech itself," whereas "the first kind depends on the personal character of the speaker, and the second on putting the audience in a

---

*This paper was presented on 15 June 1990 at the International Conference, "Science and Rhetoric," and was prepared in response to the invitation from Professor Marcello Pera, whose letter said in part: "How can scientists, during a theory change, convert their community to a new theory or way of seeing the world? We take rhetoric as the art of persuasive argumentation; we thus aim at debating its role, nature, limits as well as efficacy. . . ." He also proposed that I use twentieth-century cases, such as the theories of Lorentz and Einstein.

certain [right] frame of mind." Indeed, the chief rhetorical weapon is the speaker's inherent moral character:

> We believe good men more fully and more readily than others. . . . It is not true . . . that the personal goodness revealed by the speaker contributes nothing to his power of persuasion; on the contrary, his character may almost be called the most effective means of persuasion he possesses.

Science had to find the escape from this moralizing and personalizing mode of discourse and invent other means of persuasion than the probity or the stylistic ruses of the presenter. As if to underline that this self-denying ordinance is one of the criteria of demarcation of science, Robert Hooke's draft preamble to the original statutes of the Royal Society of London specifically disavowed that the scientists intended to "meddle" with "Rhetoric." And as we are meeting here on Italian soil, one should add that Grand Duke Leopold II, on consenting to the first Riunione degli Scienziati Italiani at Pisa in 1839, similarly made it a condition that one of the forbidden topics would be "eloquence."

Since about mid-seventeenth century, the writings of scientists have increasingly reflected their agreement with such admonitions. Thus Newton adopted for his *Principia* a structure that suggested parallels with that exemplary model of objectivity, Euclid's presentation of geometry, and he opened the first book of his *Opticks* with the implication that the work is free from conjecture, analogy, metaphor, hyperbole, or any other such device that might be identified with the rhetorician's craft. Rather, Newton says, "My Design in this Book is not to explain the Properties of Light by Hypotheses, but to propose and prove them by Reason and Experiments."[2]

The well-tested machinery of logic and analysis, the direct evidence of the phenomena—who can resist these? Who would need more? Newton and the scientists who came after liked to be considered little more than conduits through which the book of nature spoke directly, across the great divide between the independent, outer world of phenomena and the subjective, inner world of the observer. But because they are in consonance with the "Tenor and Course of Nature,"[3] their reports are free from the vagaries and imitations of mere humans. In Alexander von Humboldt's phrase, they should be the results of observation, stripped of all "charms of fancy." Or at least, as Louis Pasteur advised his students—and as is current practice in any science journal—"Make it look inevitable."

Here indeed there does reveal itself a connection with the final aim of the old rhetoric. For as Aristotle noted, the most desirable of the various propositions of rhetoric is the "infallible kind," the "complete proof" (Τεκμήριον): "When people think that what they have said cannot be refuted, they think they are bringing forward a 'complete proof,' meaning that the matter has now been demonstrated and completed."[4] Thus alerted, we now remember that a number of recent investigations by historians of science have shown that at least *before* a work has ripened into publication, during its nascent period, traditional rhetorical elements, such as conjecture, analogy, metaphor, and even the willing suspension of disbelief can be powerful aides to the individual scientist's imagination.[5] Therefore it is reasonable to ask whether some of the dramatic repertoire is not, after all, used— and perhaps even necessary—in the resulting publication also.

Indeed, I shall propose here and try to make persuasive by illustration a view different from and complementary to the usual way of reading an historic scientific paper. It is this: The publication is not only the author's account of the outcome of the struggle with nature's secrets—which is the publication's main purpose and chief strength, hence the scientist's preferred interpretation—but it may also be read as the record of a discourse among several "Actors," whose interplay shapes the publication. And as we shall see, in that respect it is analogous to the script of a play in which a number of characters appear, each of whom is essential to the total dramatic result.[6]

In using the word "complementary," I stress that I am not proposing that we may or even can choose between these two ways of reading. The second will not detract in any way from the achievement intended by the first. We shall simply be looking at the presentations of scientists not chiefly from the viewpoint of their properly intended prime audience, but as it were orthogonally, as seen from the wings. However, we must not expect that the existing published scientific work will make it any simpler to discern its internal rhetoric than it has been to derive from it the original motivation or the actual steps that led to the final result. As Peter Medawar put it in a famous passage,

> What scientists *do* has never been the subject of a scientific, that is, an ethological inquiry. . . . It is no use looking to the scientific 'papers,' for they not merely conceal but actively misrepresent the reasoning that goes into the work they describe. . . . Only unstudied evidence will do—and that means listening at the keyhole.[7]

Indeed, rare is the scientist who helps the historian or philosopher of science to penetrate beyond the mask of inevitability, to witness what Einstein called "the personal struggle," to glimpse the various influences—biographic, thematic, institutional, cultural, etc.—that gave birth to a publication.

We cannot expect otherwise, for there are good sociological reasons for that neglect and impatience. The very institutions of science, the selection and training of young scientists, and the internalized image of science are all designed to minimize attention to the personal activity involved in publication. Indeed, the success of science as an intersubjective, consensual, sharable activity is connected with the habit of silence in research publications about the individual personal struggle. Hence the useful fiction that science takes place in a two-dimensional plane bounded by the phenomenic axis and the analytic axis, rather than in a three-dimensional manifold that includes the thematic dimension.[8] Moreover, the apparent contradiction between the sometimes illogical-seeming nature of actual discovery and the logical nature of well-developed physical concepts is being perceived by some scientists and philosophers as a threat to the very foundations of science and to rationality itself. (The vogue to attempt, by a "rational reconstruction" of a specific case, to demonstrate how a scientific work should have been done seems to have been so motivated.)

Still, we shall learn how to read with minutest attention what a scientific author says or does not say, look also for the "unstudied evidence," and instead of settling only for the surface-reading that the publication invites, peer also behind the mask of inevitability. Works of literary or political intent have been subjected to an analysis of rhetorical elements for over two and a half millennia. Now we shall begin to distinguish the corresponding elements in the discourse of and about science: in the nascent phase during which the scientists weigh the persuasiveness of their ideas to themselves; in their published results; in the debates about these; in biographical and autobiographical writings of scientists; in scientific textbooks; and also in the uses made of scientific findings in controversies—a second-order phenomenon, i.e., rhetoric about rhetoric.

II. Rhetoric of Assertion vs. Rhetoric of Appropriation/Rejection

Comparing a scientific paper with the various responses to it makes it evidence that, to begin with, one must distinguish between a proactive rhetoric of assertion and a Reactive Rhetoric of Appropriation/Rejection. The first of these expresses that about which a scientist has convinced himself or herself, and hopes to persuade others of, when writing the publishable version of the work. The second characterizes the responses to it by contemporaries and later readers—responses that, we should note, are, however, shaped in turn by the responders' own commitments to their own rhetoric of assertion. The success or refusal of recognition, or its delay, as well as misplaced reinterpretation even by those who thought of themselves as converts, can thus be understood in terms of a match or mismatch between key elements in each of these two types of rhetoric.

Foremost among these key elements in many cases in the history of science are thematic commitments: those of the originator and those of the critics or opponents or would-be disciples. Since thematic commitments are not always consciously held, we are therefore often forced into a quasi-archeological task: to dig below the visible landscape of a controversy in order to find the usually invisible but highly motivating matches, mismatches, and clashes between the respective sets of themata that have been adopted by the various participants—and not only of the individual themata, but also of constellations of them that define the locally held scientific world pictures. Such correspondences and conflicts can be considered as interactions among contesting claimants in what Michel Foucault has termed "rhetorical space."[9]

Good examples for our study come from those two classic papers that, more than most others, opened the path and set the style of physical science in our century. I shall deal first with illustrative comments on Niels Bohr's seminal paper, "On the Constitution of Atoms and Molecules." Published in three parts beginning in July 1913, its main result, the working picture of the nuclear atom with its orbiting electrons, including its spectra and some indication of its chemical properties, has long been familiar to the point of banality. A physicist of today will agree with Emilio Segrè's assessment: "The sophisticated reader will admire the dexterity with which Bohr sails across a sea full of treacherous shoals and lands safely. . . . "[10]

That is, of course, how it looks to those who have been brought up on Bohr's model response. But the more immediate response was captured by Leon Rosenfeld. In his introduction to the reprint of Bohr's 1913 papers, he wrote:

> The daring (not to say scandalous) character of Bohr's quantum postulate cannot be stressed too strongly: that the frequency of a radiation emitted or absorbed by an atom did not coincide with any frequency of its internal motion must have appeared to most contemporary physicists well-nigh unthinkable. Bohr was fully conscious of this most heretical feature of his considerations: he mentions it with due emphasis in his paper, and soon after, in a letter to S. B. McLaren (1 September 1913), he writes: 'In the necessity of the new assumptions I think that we agree; but do you think such horrid assumptions, as I have used, necessary? For the moment I am inclined to most radical ideas and do consider the application of the mechanics as of only formal validity.'[11]

Indeed, if one carefully reads Bohr's paper (in volume 26 of *The Philosophical Magazine*), especially Part I, finished in haste in less than three months in early 1913, it becomes clear why it had initially such a mixed reception and why Bohr, in the interviews near the end of his life, expressed some regrets about having published it in that form. To compare the rhetoric of assertion with that of appropriation/rejection, to see how differently Bohr's work appeared to the young man himself and to some of the lions, we can also make use of fairly reliable accounts of "unstudied," spontaneous, spoken comment, of which in this case there happily exist a good supply.

Abraham Pais has published a collection of typical reactions under the heading, "It was the epoch of belief, it was the epoch of incredulity," [12] though there was at first far more of the latter. A few, notably Einstein, Debye, and Jeans, were fully receptive. But that was a distinct minority view. Thus Otto Stern told Pais that not long after the publication of Bohr's papers, Stern and Max von Laue, while on an excursion on the Uetliberg outside Zurich, swore what they called a solemn Uetli Oath: "If that crazy model of Bohr turned out to be right, they would leave physics." [13] Lord Rayleigh with lofty simplicity said of the paper, "It does not suit me." J. J. Thomson's obstinate objection to Bohr's conception was palpable in most of his writings on the atom from 1913 to 1936. H. A. Lorentz's leniency was clear, but it had its limits. As *Nature* reported in its account of the first meeting in Britain at which Bohr spoke about his atom, Lorentz (who had already objected earlier that "the individual existence of quanta in the aether is impossible") intervened to ask "how the Bohr atom was mechanically accounted for," and Bohr had to acknowledge "that this part of his theory was not complete, but . . . some sort of scheme of the kind was necessary." [14]

Bohr himself noted later,

> When my first paper came out, it was actually objected to in Göttingen. There was no interest for it, and, as I told you, there was even a general consent that it was a very sad thing that the literature about the spectra should be contaminated by a paper of that kind. The paper was just a playing around with numbers and there was nothing in it. . . . It was clear that that was the general consent. . . . Because at first there actually was nothing. And that's what we'll come to. But now the question is, how was it presented?" [15]

That was indeed the question. In a preview of his work, Bohr had warned Rutherford in 1912 that he, Bohr, would have to adopt an hypothesis "for which there will be given no attempt at a mechanical foundation (*as it seems hopeless*)." [16] But when Rutherford actually saw the manuscript, he had to write to Bohr on 20 March 1913,

> the mixture of Planck's ideas with the old mechanics [Bohr himself had characterized it as "the delicate question of the simultaneous use" in a letter of 6 March 1913] makes it very difficult to form a physical idea of what is the basis of it all. . . . How does the electron decide what frequency it is going to vibrate at when it passes from one stationary state to another?

A fair question—it took until 1917 for Einstein to show a way.[17]

What most concerned many of Bohr's readers—brought up on atom models, such as Thomson's, that were considered "mechanically accounted for"—when forced to decide on appropriation or rejection, was not only Bohr's presentation, a rhetoric of assertion in which he rather cavalierly mixed classical and quantum physics, but also his introduction into his atom of the thema of discontinuity as well as that of probabilism rather than Newtonian causality—antithemata with respect to the classical foundations. These were threatened at the time also from other directions. Returning from the 1911 Solvay Conference, the first summit meeting on quantum physics, James Jeans had baldly stated what to many was an ominous advent in the thematic base of physics: "The keynote of the old mechanics was continuity, *natura non facit saltus*. The keynote of the new mechanics is discontinuity."[18]

But Jeans was far more ready for this profound change than many others. Eddington said of him that he was the only one in England who had been converted to quantum physics by the Solvay Conference. Henri Poincaré, returning from the same meeting, spoke for the large majority when he concluded wistfully in the last year of his life:

> The old theories, which seemed until recently able to account for all known phenomena, have recently met with an unexpected check. . . . A hypothesis has been suggested by M. Planck, but so strange a hypothesis that every possible means must be sought for escaping it. The search has revealed no escape so far. . . . Is discontinuity destined to reign over the physical universe, and will its triumph be final?[19]

Unlike so many of his elders, the 27-year-old Niels Bohr had built up no equity in the themata of the older physics. He was young enough to have encountered the existence of quantum ideas from his student days on. Moreover, in working on his doctoral dissertation on the electron theory of metals, which he had just completed, he understood more clearly than his own examiners that the classical conceptions were simply incapable of dealing sufficiently with the situation, for example, with specific heats, or the high-frequency portion of black-body radiation, or the magnetic properties of matter.

Thus, to understand the argument by which the author of a work has convinced *himself*, one must look for its roots that may have already appeared in his previous work. Bohr's 1913 paper is a point on a developing trajectory of personal science (S_1) that intersects upon publication in July 1913 with public science (S_2). On the earlier part of S_1, we find not only Bohr's doctoral thesis but his abortive discussions with J. J. Thomson during Bohr's stay in Cambridge, and

Bohr's productive work on alpha particle scattering at Rutherford's laboratory in Manchester; their traces can be found on the first pages of the July 1913 paper.

III. Not One Actor but (at Least) Two

We generalize this point in the following proposition:

I. *A scientist's current work is likely to be the continuation of a soliloquy that has its origins in his earlier work.*

A second proposition follows as if by symmetry:

II. *In the scientist's current work one may discern evidences of the direction that his future work is likely to take.*

To add to the illustrations already given for proposition I, we may note that Bohr's courage in July 1913 is a consequence of his earlier radicalization. Rutherford's nuclear model of the atom was discovered quite unexpectedly at the end of 1910 and published in 1911. It, too, was at first widely disbelieved, and Rutherford did not insist on it himself (as indicated by his silence about it at the 1911 Solvay Congress). But its implications were enormous and were perhaps best caught in the artist Kandinsky's outburst that now that the old atom had been destroyed, the whole existing world order was destroyed, and so a new beginning was possible.[20] To Rutherford's young collaborators in Manchester, especially to Bohr who had fled there from Cambridge and its resistance to new ideas, Rutherford's discovery of the concentration of the atom's mass had revealed the crucial flaw in the then reigning model of the atom (primarily J. J. Thomson's), even though that model had, among other useful features, yielded a plausible explanation for the size of the atom and for multiple-scattering data.

However, in Rutherford's atom model, no one knew any longer what to do with the electrons around the nucleus. Thomson thought of that as "a very great calamity";[21] but when Bohr was asked, "Were you the only one who responded well to it [the Rutherford atom]?" He replied, "Yes, but you see I did not even 'respond' to it. I just believed it."[22] He had evidently been ready for it after his unsatisfactory struggle with the classically based atom during his dissertation work: "Now it was clear, and that was *the* point in the Rutherford atom, that we had something from which we could not proceed at all in any other way than by a radical change."[23] Or, as he had put it in his July 1912 "Memorandum" for Rutherford, the stability of the electrons' configuration had to "be treated from a quite different point of view."[24]

The direction in which to seek salvation was clear. For some years, Planck's quantum of action h had been the tool for understanding black-body radiation, and it promised to do the same for specific heats. It had a magic about it, at least for young people ready to risk it. (As E. C. Kemble, who initiated quantum physics research in the United States in his twenties, recalled as his own motivation: "Anything with quantum in it, with h in it, was exciting."[25]) It was, as so often in the history of science, a matter of being ready to embrace the new themata. Even those features of Bohr's atom, which to others eventually were the most persuasive—e.g., the correct prediction of spectral lines and the derivation of the value for Rydberg's constant—were not essential for convincing Bohr himself; for we know now that he stumbled on these aspects only at the last minute, in early 1913, when the main parts of his paper had been fixed.

* * *

It is also easy to illustrate proposition II. Thus a striking feature of Bohr's 1913 paper is its first demonstration of the dialectic nature of Bohr's thinking, which then suffuses all his later works, including especially that on complementarity. His radicalization had not forced him, as it might have others, to abandon entirely the old, mechanistic conception. On the contrary, he held that "by analogy" to what is known for other problems, it seemed to him legitimate to continue to use the old mechanics *side by side* with the new quantum physics, often within the same paragraph of the 1913 paper. This is just what Rutherford had found most puzzling. But it is at the heart of Bohr's daring proposal in 1913 of what he named later the "correspondence point-of-view," which in turn, from 1927 on, burgeoned into his "complementarity argument."

We can now summarize this segment: To a greater or lesser degree, a publication can be read as the extrapolation from the author's past, as well as the staging area for a future expedition. To put it differently, in studying the Rhetoric of Assertion of an author in a given work, we discern *that he disaggregates into two Actors, engaged in two different soliloquies on the same stage*.

Actor I is engaged in an internal dialogue with his own recent or more distant past work, out of which the new work is growing. Actor 2 has begun to engage in thoughts that will not come to full fruition for some time in the future. The author's production results in good part from both soliloquies and receives different characteristics from each: on one side, conviction from past difficulties being now conquered; on the other side, conviction from the attractiveness of further successes that perhaps only dimly but tantalizingly beckon—especially in Bohr's case, the new thema of complementarity; the hope for a greater unification of understanding both chemical and physical properties of matter through his new atom; and the feeling that something wonderful looms beyond. Thus in his letter to G. Hevesy, Bohr writes on 7 February 1913,

> . . . I don't speak of the results which I mean that I can obtain by help
> of my poor means, but only of the point of view—and the hope to and

belief in a future (perhaps very soon) enormous and unexpected??
development of our understanding—which I have been led to by
considerations as those above.[26]

It is a near paraphrase of Galileo's prophecy, at the end of Day Three in his *Two
New Sciences*, that "the principles which are set forth in this little treatise will . . .
lead to many another more remarkable result." In this way, while Actor 1 is
animated by the satisfaction of recent difficulties surmounted, Actor 2 is pulled
forward by attraction to the greater goal on his agenda. Moreover, one of the most
important long-term functions of a seminal paper in science is plainly rhetorical:
that its readers come to share the author's excitement, his sense that new vistas are
being opened, that new questions can be raised and perhaps answered. (The
chemist Dudley Herschbach has christened it "the spiritual effect" of good new
science.)

* * *

But as we also have begun to note, the two Actors are by no means alone on the
stage defined by the text of the paper. Each carries on his monologue in the
imagined presence of his important colleagues. The published paper bears witness
to that: One can unravel which of the two Actors is speaking a line, and against the
background of which other imagined supporter or opponent that line is composed.
(By no means all of these will be identified in the text by name or in the notes.)
Thus Bohr's paper in its first paragraph is the acknowledgment by an acolyte of
"Professor Rutherford" (who also served as the identified communicator of the
paper to the journal), of the motivating power of Rutherford's recently discovered
nuclear atom. The next two paragraphs are a continuation of it, with the addition of
a cautious acknowledgment of the power still exercised by the commanding ghost
of "Sir J. J. Thomson." In the fourth paragraph, we see Bohr accepting the promise
of the revolution Planck had introduced in 1900 (much against his own will). And
only then, in the last half sentence of that paragraph, in a throw-away line, the first
evidence of Bohr's own ideas: a remarkable feat of confident intuition, made
almost incomprehensible by the failure of the young author to articulate his own
voice in that distinguished company.[27]

By page four, Bohr introduces the strange idea that the frequency of the
radiation emitted in binding the electron to the atom is "equal to half the frequency
of revolution of the electron in its final orbit." A typical early reaction to that was
the term "a crazy stunt"—but here we have again the emergence of Actor 2 on the
stage, presenting an apparently unsupported argument that will develop later into
Bohr's treasured correspondence argument by which he tries to hold on to both
classical physics and quantum physics.

* * *

In this way, a paper can be resolved paragraph by paragraph into the main rhetorical components in the assertion stage, e.g., into the various parts that are carried by different Actors. Moreover, one can also differentiate between the rhetorical components in the subsequent stages of appropriation or rejection— which in the case of Bohr's atom was particularly turbulent for the first years among physicists in the United States, who were puzzled whether to regard Bohr's two-dimensional atom model as a discovery, an analogy to the three-dimensional nature of matter, or a powerful metaphor.[28] But instead of pursuing this further for this particular case, and to indicate the universality of the role of rhetoric despite great individual differences, I turn now to another seminal paper in the history of early twentieth-century physics.

IV. *Relativity: Its Publics and Its Authors*

That case is analogous to the last one chiefly in one respect: Albert Einstein's formulation of what is now called special relativity has also become so familiar to us that one may say, as he did about Ernst Mach's ideas, that one has imbibed it with one's mother's milk. Eventually, relativity theory became one of the "charismatic" activities, to use the terminology of Joseph Ben David. Therefore it takes an act of serious will to free oneself from an ahistorical view about Einstein's claims, as they were launched in a quick series of communications, starting with the publication of the first paper on 26 September 1905, "On the Electrodynamics of Moving Bodies." From the perspective of rhetoric, it was almost calculated to be off-putting to the typical reader of 1905. Indeed, as Einstein had predicted in one of his letters, in terms of the immediate reception by the large scientific community, this work could be regarded as a failure. Approval was certain only from the few personal friends of this unknown and sociologically "marginal" man, fellow marginals, such as Michel Besso, Joseph Sauter, Marcel Grossmann, and Conrad Habicht. While Bohr's paper showed from the first sentence that he was conscious of moving, as indeed he did, in Olympian company, Einstein's emanates the sense that the young author is unused or unwilling to address himself properly to his "betters" (as indeed was also the case).

A fair understanding of Einstein's formulation grew among major physicists only slowly during the first few years. The Rhetoric of Appropriation/Rejection was heavily weighed to the latter. And even those who one by one were converted, in almost all cases interpreted the main point of Einstein's work in a significantly different manner than he himself had intended. It is reasonable to say that it took six years, with the appearance of Max von Laue's first textbook on relativity in

1911, for an irreversible change in the unfavorable balance to be signalled; and some, including H. A. Lorentz, did not make their peace with Einstein's relativity to the end of their days. The one great exception in all this, as we shall see, was Max Planck, whom Einstein himself regarded as his first and crucially important champion among the elite. And even there, when it came to the extension into general relativity some eight years later, Einstein complained in one of his letters to Ernst Mach that Planck's "stance to my theory is also one of refusal." [29]

As for the other physicists whose work Einstein had studied and admired, such as Wilhelm Wien and Henri Poincaré, he surely must have hoped for some early and real understanding of what he was trying to do. But on that score, Einstein was to be completely disappointed; and Mach, after early expressions of brief, diplomatic, and cautious words of encouragement, turned against relativity when he began to recognize what the program of relativity was and what it demanded.[30] Hermann Minkowski's enthusiastic embrace in 1908 of relativity theory—in his own reinterpretation—left Einstein himself at first quite cold. For the next few years, younger scientists and philosophers, such as Friedrich Adler in Switzerland, Joseph Petzoldt in Germany, Paul Langevin in France, and Richard C. Tolman and Gilbert N. Lewis in the United States, began to adopt relativity for their own purposes. But again, more often than not, they initially misunderstood the main point. The same pattern of cases of either appropriation by misinterpretation or outright rejection continued in some circles for decades.

These various responses to a theory that now seems so clear to scientists call for explanation. In such matters one does not expect to find just one or two mechanisms, and not all of them need have been clear to the participants themselves. But even a brief list of such reasons must contain facts, such as these: that Einstein's first paper on relativity theory had even greater ambitions than those openly stated; that it was complex and strangely construed, as seen by those habituated in the then current style of physics—in effect a violation of the contemporary Rhetoric of Assertion in physics—whereas for us, inheritors of much of Einstein's way of thinking and arguing, the paper makes far fewer demands; that Einstein's proposals were really not *necessary* for a physicist in 1905 because what William James would have called the theory's "cash value" for contemporaries did not seem to be superior to those derived from, e.g., Lorentz's quite differently based and quite successful theory; that it asked for large conceptual sacrifices to be made (such as abandoning the absolutes of time and simultaneity, and the ether) in return for the relief from major pains that only the unknown young author seemed to feel.

To top it off, the paper in its published form, written hastily after seven years of reflection within five or six busy weeks, had—in addition to errors that soon had to be corrected—[31]a cavalier air about it. That is indicated, for example, by its unusual failure to have any bibliographic references, and by its resistance to demonstrate clearly some of its own favorable points and implications, such as that what are still called the Lorentz transformation equations were now derivable from Einstein's postulates and thus did not have to be introduced in a manner both Lorentz and Einstein considered ad hoc. Recalling Aristotle's three kinds of "modes of persuasion" necessary for the good rhetorician—exhibiting the good personal character of the speaker, putting the audience in the right frame of mind, and providing a proof through the speech itself—we note that none of these seemed to weigh on Einstein. If anything, he seemed to be paying as little attention to them as possible.

I do not know which of these or other "flaws" were on Einstein's mind when he himself in the 1940s came to express displeasure with his 1905 paper. The occasion was the following, as related to me by his long-time secretary, Helen Dukas: Einstein had been asked to donate the manuscript of his 1905 paper to a fund raising drive on behalf of United States government war bonds. Because he had not kept that manuscript, he decided to make a new, handwritten copy from the published version. (It actually fetched a huge sum for the government in the auction and now resides in the Library of Congress.) To speed the work of copying, Einstein had Helen Dukas dictate the paper to him. She told me that instead of following her dictation faithfully, he repeatedly objected that he "could have said it much better," and indeed now intended to do so. She had to plead with him constantly to keep him from improving on his old work.[32]

At any rate, if one analyzes the 1905 relativity paper with care, line by line,[33] one can again discern throughout the existence of two Actors engaged in their different monologues, one with his past, the other with his future. A look at the first few lines will suffice here to make the point. Thus Einstein's first paragraph is centered on a retrospective reflection of Actor 1 upon Einstein's own early struggle with classical electrodynamics, as he experienced it in his student years, e.g., in reading August Föppl's text *Einführung in die Maxwellsche Theorie der Elektrizität*, 1894 (which in turn had acknowledged epistemological debts, particularly to Kirchhoff, Hertz, and Mach). The construction of Einstein's initial *Problemstellung* in the 1905 paper is completely parallel to Föppl's fifth main section, and includes especially the Faraday experiment referred to by both. It played, as Einstein repeatedly noted, "a leading role" in "the construction of the special relativity theory."[34]

In the second paragraph, we continue to hear echoes of the concerns of Einstein's earlier self, including the *Gedanken*-experiment at about age 16 and the abortive plans for actual experiments, made while a student at the university. But we also begin to discern Actor 2 in the decision to remove the barriers separating the laws of physics, starting with those between mechanics and electrodynamics.

For it was the most enduring passion of Einstein, from his earliest years as a scientist to the end, to pursue what he called (in a letter to W. deSitter) "*my need to generalize*" ("*mein Verallgemeinerungsbedürfnis*").

That need appeared already while he composed his first published paper (1901) on the unlikely subject of capillarity. He writes to his friend Marcel Grossmann (letter of April 15, 1901) that he is trying there to bridge the molecular forces and Newtonian forces at a distance, and he bursts out: "It is a magnificent feeling to recognize the unity [*Einheitlichkeit*] of a complex of phenomena which to direct observation appear to be quite separate things." Similarly, he reported in a manuscript written about 1920 that in writing the 1905 paper he had found the contemporary interpretation of the Faraday experiment "unbearable" because it regarded as "two fundamentally different cases" what he felt needed to be subsumed under one more general case.[35] And in virtually each of the other papers preceding *or* following upon the relativity paper of 1905, we find that the appeal of generalizing takes over and becomes a directive for research. We know now that even while working on special relativity Einstein felt it to be too limited, and hence decided to extend the postulate of relativity to nonuniformly moving coordinate systems.

To put it more starkly: When Einstein begins his work, he is aware that physicists are deeply divided between the program and claims of the mechanistic world picture and the electromagnetic world picture. Already in his third paper (written in 1902) on extending Boltzmann's ideas in thermodynamics and statistical mechanics, he joins the battle head-on by testing some limits of what he calls there the "*mechanische Weltbild*." By the time he is writing the relativity paper, he has seen that neither the mechanistic nor the electromagnetic world picture by itself suffices, e.g., in dealing simply with fluctuation phenomena. Nor would a victory for one or the other have satisfied him; as he said later, it would, for example, leave us with "two types of conceptual elements, on the one hand material points with forces between them, and, on the other hand, the continuous fields, ... an intermediate state of physics without a uniform basis for the entirety."[36] Without some awareness of these agendas of Actor 2 for the future, Einstein's paper must have been far more puzzling to his contemporaries than it is to us who know how it all turned out.

The motivating words "Weltbild" or "Verallgemeinerung" or "uniform basis for the entirety," of course, do not appear anywhere in the 1905 paper. But my point here is emphatically that just as there is a danger of blindly reading ahistorical elements back into earlier work, there is equally a danger to being blind to the forward thrust that may silently underlie the program of research at a particular time. One will not understand Actor 1, speaking at time *t*, without having made a detailed historical study of what preceded *t*. But one will not even properly hear Actor 2 unless one has studied what followed after t. Many excessively "internalistic" studies of a scientific publication have failed to catch the spirit of the work for that reason.

V. The Stage Fills

Now we turn our gaze more directly on the group of Actors who in our metaphor constitute the dramatic personae embedded in the text, although the scientist-author usually will claim to be giving us merely access to nature itself as revealed directly through "Reason and Experiments." In the case of Bohr's paper, we saw him turning to Rutherford, Thomson, and Planck by summarizing what he perceived to be correct and important or incorrect and incomplete about their prior work in this field. We could have added others; for example, Bohr has a lively though one-sided "conversation" with J. W. Nicholson (on pp. 6—7, 15, 23—24 of Part I of his paper, and more in Part II) about Nicholson's doomed theory of line emission spectra.

For the purpose of such "conversations," those other scientists-colleagues in the case are brought on the stage in the author's script explicitly or implicitly. But, of course, they are presented to us on the author's terms—their voices and proposals are adjusted or interpreted to serve the script. While Bohr was meticulously fair, and usually no distortion is intended, occasionally we do hear later from one of them in their own voice, when they take exception to what they perceive to have been a misunderstanding of their true position. At any rate, the Rutherford, Thomson, Planck, or Nicholson of whom we learn in Bohr's paper cannot, with the best will in the world, be considered fully representative of the originals. On this stage, alongside the two "Bohrs," they are Actors 3, 4, 5, 6 . . . , speaking lines that their corresponding models might not have thought of.

The same considerations apply to the main stage filled by the characters in Einstein's paper. There we encounter, in addition to Einstein serving as both Actors 1 and 2, also H. A. Lorentz—but only a mere fragment of the Lorentz we know to have existed in 1905, for Einstein had not yet read Lorentz's key paper of 1904,[37] and was not convinced by those publications that he had read. We have already met Einstein's Föppl, one of several characters not mentioned by name in Einstein's paper. Ernst Mach is also not named; but a facsimile of him presides so visibly over the section "Kinematical Part" of Einstein's paper that a whole generation of positivistically inclined scientists and philosophers (from Petzoldt to Heisenberg) was misled to think of the whole paper as primarily a triumph of positivism. Other partially recognizable but anonymous Actors who make appearances bear fainter likenesses—Helmholtz, Hertz, Boltzmann, Wien, Abraham, and, the faintest voice of all, David Hume. There may be others. For example, because we lack here the wealth of drafts and letters that we have for Bohr and his circle, written during the crucial period of composing his paper, we do not know which passages in Einstein's paper may refer directly to his conversations with friends, such as Besso.[38]

Einstein's representation of "Lorentz" is a particularly interesting character in his own right, as indeed is Lorentz's "Einstein" in Lorentz's later publications. After their first meeting in 1911, Einstein came to admire and even love Lorentz as a superb physicist and a remarkable person; and Lorentz's fondness for Einstein was also very deep when they came to know each other. But just as Lorentz never accepted relativity fully, Einstein had not much patience with Lorentz's approach to electrodynamics.

It is therefore highly ironic and appropriate for a study of rhetoric in science that during the early years the very different research programs of both men were widely subsumed in the literature under the joint name "Lorentz-Einstein." That fiction is worth more than a brief glance. One of the first to use it was Walter Kaufmann in early 1906, in the first article in the *Annalen der Physik* to respond to Einstein's 1905 paper—by putting "the Lorentz-Einsteinian fundamental assumption" to the test.[39] As we shall see in more detail below, Planck thereupon took up the cudgel on behalf of Einstein's relativity. But he too began his talk[40] with the line that "recently H. A. Lorentz, and in more generalized form Einstein, [had] introduced the Principle of Relativity"; and soon thereafter,[41] Planck, too, used the term "Lorentz-Einsteinian" theory.

Of course, there is a sense in which one may read Lorentz's 1904 paper and Einstein's 1905 work as operationally "equivalent"—another potent term that will deserve more than passing comment. Both theories used almost the same transformation equations and thus allowed effectively the same observable results to be derived with respect to experiments of interest at the time. Apart from that, however, the two theories were at opposite poles in every respect—in terms of their genesis, their physical and philosophical underpinnings, their respective assumptions, including the thematic ones, and their implicit further goals. In short, they were the products of quite different world views.

For example, as Lorentz's book of 1895[42] and the structure of his 1904 papers show, his work was driven largely by the strategy of patching up a theory of the electron that had been battered by puzzling recent experiments, whereas Einstein's paper was, as he stressed over and over again, motivated by the desire to build a coherent physics "by the discovery of a universal formal principle" on the model of thermodynamics[43] and helped by his reinterpretation of old and well-known first-order experiments (Faraday's stellar aberration and Fizeau's measurements of light propagation in moving water ["They were enough," as Einstein told R. S. Shankland]). Lorentz did not hesitate to continue to introduce what he himself regarded as "somewhat artificial," ad hoc auxiliary conceptions as needed,[44] even after being scolded for it by Poincaré. And he freely confessed in 1912 that his theory of 1904, built around the model of a deformable, mechanically unstable electron, exhibited "clumsiness" and incompleteness, while only Einstein's provided "a general, strictly and exactly valid law."[45] Moreover, Lorentz's was essentially a physics of a particle, the electron, whereas Einstein's was a physics of any event in space and time. The ether provided obviously yet another demarcation criterion between the two, Lorentz's physics being firmly based on it to the end, while Einstein had dismissed it in an early passage with a casual wave of the hand.

We therefore do not find it surprising that their respective world pictures are entirely different also: on Lorentz's side, the best representation of the electromagnetic *Weltbild* available at the time; on Einstein's side, a new one that demanded applicability across all fields of physics, as well as the elimination wherever possible of asymmetries, ad hoc hypotheses, and redundancies (the existence of which Einstein found "unbearable"), no matter what cost it would entail in terms of resulting conceptual rearrangement. But such basic differences in the underlying world pictures were slow to be recognized,[46] and the long persistence of the term "Lorentz-Einstein" was an indicator of it.

VI. *An Experiment in the Rhetoric of Appropriation/Rejection*

Having watched the main stage fill with agents implied in the rhetorical space of Einstein's own paper of 1905, we can now visit a *side stage*, on which the "real" versions of the characters carried out their acts of appropriation or rejection of what Einstein had to offer, immediately after his publication.

Here we are fortunate in that there took place a public encounter of opposites that may serve as an "experiment" in the Rhetoric of Appropriation/Rejection and reveals some of its fine structure. It was in fact initiated by the publication of Walter Kaufmann's papers of 1905 and 1906, claiming to give the empirical test data that would crucially decide between the current theories.[47] For our purposes we need only cite the results that this distinguished, Göttingen-based physicist himself put near the start of this major experimental examination (finished on 1 January 1906) of Einstein's 1905 work. Kaufmann wrote in italics: "I anticipate right here that the . . . measurement results are not compatible with the Lorentz-Einstein fundamental assumptions." Further on, again in italics, he declared those assumptions "a failure." For good measure, Kaufmann pronounced his data to favor a recent, much more limited theory by Max Abraham.[48]

Moreover, in an addendum less than four months later, Kaufmann implied that if one wanted to distinguish between these two discredited approaches, despite the equivalence of their derivable predictions of empirical facts, Lorentz's had one advantage over Einstein's. For the inductivist methodology of Lorentz had yielded in his case the proposed "independence of all observed phenomena from a uniform translation" as an "end result," whereas Einstein had merely proposed it initially "as a postulate, at the apex," achieving thereby the same system of equations "through pure mathematics."[49]

That was little comfort for Lorentz. Most respectful as always of the work of experimenters, this great theoretician saw his labors of well over a decade suddenly destroyed even by Kaufmann's preliminary (1905) results. He seemed devastated, writing to his friend and fellow theoretician Poincaré on 8 March 1906: "Unfortunately my hypothesis . . . is in contradiction with Kaufmann's results, and I must abandon it. I am thus at the end of my Latin." He appealed to Poincaré for help. But none came; instead, Poincaré noted that the "entire theory" may well be threatened by Kaufmann's results.[50]

Einstein's response to Kaufmann has also been noted before; it was completely different, not least perhaps because Einstein had a healthy skepticism about the claims of new experiments. At first, Einstein ignored the results, and it took an appeal from Johannes Stark in 1907 to survey the state of the relativity theory to move him. In brief, Einstein indicated that his theory had been taken too narrowly, as only a contribution to electrodynamics; that it was possible that the data from such a difficult experiment as Kaufmann's could be in sufficient agreement with his own theory after all; and that systematic errors in Kaufmann's data seemed likely.

But above all, Einstein's intuition told him something to which others were not alert: the data which seemed so conclusive may have been faulty *because* they favored theories, such as Abraham's and Bucherer's, which applied to a rather small region of physics compared to Einstein's own:

> In my opinion both theories have a rather small probability, because
> their fundamental assumptions concerning the mass of moving elec-
> trons are not explainable in terms of theoretical systems which embrace
> a greater complex of phenomena.[51]

In the meantime, there had recently taken place a most revealing debate concerning the grounds for believing in any of the theories in the absence of incontrovertible empirical evidence. It began with Max Planck's quick response to Kaufmann's publication. At 48 years of age one of the most distinguished physicists in the world, and well on his way to becoming the dour dean of German physics, Planck showed that he was sensitive to the deepest meaning of Einstein's 1905 paper. In a brief talk of 23 March 1906, he declared that if the "Principle of Relativity" (as he called the theory at first) were "borne out, it will be a grand simplification of all problems in the electrodynamics of moving bodies. . . . " He added that a thought of such "simplicity and generality" deserved, even in the face of Kaufmann's claimed disproof, to be subjected to more than just one test; and if the idea then did turn out to have been defective, it should nevertheless be taken *ad absurdum*, and its consequences examined.[52]

A few months later, Planck undertook a long, detailed reexamination of Kaufmann's recent result of experiments on the deflection of beta rays, which "for different electrodynamic theories is so-to-speak a question of life or death."[53] He then recast the theoretical base of Kaufmann's experiment more thoroughly than Kaufmann himself had done. While also revealing the considerable number of assumptions that Kaufmann needed (e.g., field homogeneity), Planck compared the reported observations with the expected values that can be calculated on the basis of "those two theories which so far have been most developed," i.e., that of Max Abraham (1903), "and the Lorentz-Einsteinian [as noted, Planck also made use of the term] in which the Principle of Relativity has full validity."

Even in this reexamination, Planck found the "data" that Kaufmann had published to be closer to the prediction of Abraham's theory than to the "Lorentz-Einsteinian"; typically, observation yielded 0.0247 units, while the first theory predicted 0.0262 and the second 0.0273. But as if to show that "data," too, have rhetorical uses, does not see Planck in those numbers "a definite proof of the first and a disproof of the second theory." After all, the differences between the predictions from the two theories were generally smaller than the difference between Kaufmann's reported "observations" and either of the theoretical values. Hence, Planck notes, one can begin to suspect a systematic error in the experiment or in its assumptions. "There seems to be a significant defect [*Lücke*]" somewhere; hence a definite decision between the theories is at this point unwarranted. Moreover, Planck finds the whole experiment a bit misguided, for it uses fast beta-rays, whereas a better decision between the theories can be shown on theoretical grounds to be expected from the use of slower electron beams.

Happily, Planck had chosen to present these findings in a public lecture (Stuttgart, *Deutsche Naturforscherversammlung*, 19 September 1906), and after it there ensued a lively discussion that was also published.[54] For a study of the Rhetoric of Appropriation/Rejection, it is a revealing and even amusing theatrical script of its own. Walter Kaufmann rose first; he was glad to see that Planck's calculations, made on a different basis, had resulted in "identical numerical results"—only a slight exaggeration—and so gave one confidence that no errors of calculation had entered. But after all, Kaufmann had to insist, the Lorentz-Einstein (L-E) theory predictions deviated from his data throughout by 10 to 12 percent, whereas Abraham's (A) theory came to within 3 to 5 percent—also outside the error of observation, but possibly within the error from *all* sources.

Planck's response was uncharacteristically curt. In the absence of a full understanding of the error sources in addition to observation errors, it was for him quite thinkable that when such corrections eventually might be made, they would bring the data closer to the L-E theory than to its rival. A. H. Bucherer now rose to reflect in a rambling speech on how Planck's analysis affected his own theory, one similar to Abraham's, and how it might be improved. (In passing, he did make, as had rarely been done so far, a distinction between Lorentz's and Einstein's theories, both of which he believed to be flawed for different reasons; and he was the first person to adopt Planck's newly proposed term, "*Relativtheorie*," but shortly after coined the term "*Relativitätstheorie*.") However, for his labors he was rebuffed by Planck, who asked him about a "very important" test of Bucherer's theory, which Bucherer had to confess he had not yet made.

Now it was Max Abraham's turn to speak. It must be remembered that he was a brilliant physicist, whom Einstein also respected, but whose theory of the rigid electron was based on a completely different, fiercely held world picture. As von Laue and Max Born noted some years later, Abraham

> found the abstractions of Einstein disgusting in his very heart. He loved his absolute ether, his field equations, his rigid electron, as a youth loves his first passion whose memory cannot be erased by any later experience. . . . [Einstein's] plan was to him thoroughly unsympathetic.[55]

Abraham rose and asserted (to "great laughter") that since the predictions from the L-E theory deviated from Kaufmann's data twice as much as his theory did it followed that his own is "twice as good as . . . the *Relativtheorie*." He was satisfied with the result. Moreover, his own theory had the advantage that it was a "purely electromagnetic one." Even Lorentz's failed by that criterion because it assumed (as Poincaré had also found recently) the need for a term in addition to its electromagnetic energy.

Planck replied that he agreed with that fully—but so far Abraham's purely "*elektrische*" theory was only a hopeful postulate, an unachieved program. To be sure, the L-E theory "is also based on a postulate, namely that no absolute translatory motion can be discovered." So in Planck's opinion, we had here two unproved and undisproved theories. And at that crucial point, having put before his distinguished audience the choice between the two antithetical postulates, and hence between these antithetical conceptions of reality, the magisterial Planck added a few sentences that surely deserve to be a highlight of any future theory of the rhetoric of science:

> These two Postulates, it seems, cannot be united; and so it comes to this: to which Postulate [L-E or A] to give preference. *As to myself, the*

> *Lorentzian is really more congenial. [Mir is das Lorentzsche eigentlich sympathischer.]*

Under pressure when the chips were down, the inner motivation for making a choice in the absence of meaningful differences obtainable through Newton's "Reason and Experiments" became visible: it is the *feeling of sympathy, of congeniality with one world picture rather than with its opposite, a decision based on one's scientific taste.*

Having made this revelation (in which he had condensed the term for the L-E theory further, into merely Lorentz's), Planck immediately added a protective sentence: "It would be best if both fields were to be further developed, and in the end experiment provided the decision." Thereupon Arnold Sommerfeld, at 38 years of age one of the bright newer stars of physics, felt compelled to remark he could not join Planck in the "pessimistic point of view" that the decision should be delayed until experiments spoke more clearly. Leaving aside Kaufmann's results because the "extraordinary difficulties of the measurements" might well have produced deviations from the expected data that came from still unknown sources of error, Sommerfeld could make known his choice now:

> I suspect that regarding the question of Principles which Herr Planck
> has formulated, preference is given to the electrodynamic Postulate
> [i.e., the Abraham theory based on the electromagnetic world picture]
> by those under 40 years of age, and to the mechanistic-relativistic one
> [i.e., to the Einsteinian extension of the principle of relativity to all of
> physics] by those over 40 years. I prefer the electromagnetic one
> myself [laughter].

While Sommerfeld's division was not quite correct, and he soon changed allegiances, his confession also underlined that congeniality of point of view is a criterion for theory choice in science—even as rhetoricians from classical Greece on knew it to be in the three traditional areas of display: political, forensic, and ceremonial.

In an anticlimactic ending of the discussion, Kaufmann got up once more. He objected that "the epistemological worth" of the relativity postulate was small because it did not apply to systems other than inertial ones—a point that Einstein, of course, knew well, and which was propelling him toward the "generalization" in which he was to succeed shortly. Planck squashed Kaufmann in three sentences: Kaufmann had missed the main feature of the relativistic point of view—that what could not be observed in inertial systems by mechanical experiments should also be unobservable by electrodynamic ones.

Nothing had happened that *forced* anyone to change his mind, to abandon one theory together with the world picture on which it was based, and to favor adopting the opposite. Individual experimental results of a narrow sort continued to come in for some years and lent themselves to one cause or the other, depending on how robust one thought the underlying assumptions were. Thus for a few years, the scientific community found itself somewhat at a loss how to deal with two such differently based theories whose "cash values" were about the same—and both of which were still under the cloud of the Kaufmann experiment (which was not fully unmasked as defective until 1916). In another irony that must have amused Einstein, the best support for both theories for years was thought to be the fact that in their fundamentally different ways each "explained" the haunting failures to find ether drift effects.

VII. By Way of Epilogue: The Inertia of Rhetoric

When *was* it over? When did Bohr's and Einstein's "art of persuasive argumentation" succeed respectively in converting "their community to a new theory or way of seeing the world," to quote from our letter of invitation to this conference? The superiority and scope of Bohr's theory, plus the results of decisive experiments (e.g., Franck-Hertz, 1914) had made it soon irresistible. But in the case of relativity, there was no hope of having some crucial experiment decide quickly between the rival theories, with their bases in vastly different world pictures. What had to happen, as so often, was a slow process by which more and more of the visible members of the scientific community learned to hear and understand the voices on the stage for which Einstein had written the script. For example, the perceptive and well-placed physicist Wilhelm Wien, with whom Einstein had begun a correspondence in 1899, had initially published his disagreement with relativity; but by 1909 he had become persuaded by it and its world picture essentially on aesthetic grounds. He wrote:

> What speaks for it most of all, however, is the inner consistency which makes it possible to lay a foundation having no self-contradictions, one that applies to the totality of physical appearances, although thereby the customary conceptions experience a transformation. [56]

We noted that the appearance of Max von Laue's textbook of 1911, entitled significantly still *Das Relativitätsprinzip*,[57] essentially dates the first solid indication of the victory of Einstein's world conception over Lorentz's (and Abraham's); but even then von Laue had to confess that

> a really experimental decision between the theory of Lorentz and the Relativity Theory is indeed not to be gained; and that the first of these nevertheless had receded into the background is chiefly due to the fact

that, close as it comes to the Relativity Theory, yet it lacks the great simple universal principle, the possession of which lends the Relativity Theory from the start an imposing appearance.

Indeed, after Kaufmann,

very significant experiments by Bucherer [1909] and E. Hupka [1910] seemed to speak in favor of the Relativity Theory, but opinion about their power of proof is still so divided that Relativity, from that side, has not yet received unquestionably reliable support.

Von Laue added that the wealth of different phenomena encompassed by relativity theory was so vast that it was a task of the highest order to achieve an explanation of all of these by the adoption of one point of view. Thus "it is no wonder that this task reaches deeply into our whole physical *Weltbild*, and touches on the epistemological foundations of science."

It took some years more for the special relativity theory to become truly a widely accepted part of physics. That had to wait for developments far from the scope of Einstein's 1905 paper itself—foremost among them experimental successes, such as the eclipse expedition of 1919 with its test of a prediction of the general theory of relativity, and the use of relativistic calculations to explain the fine structure of spectral lines.

In the meantime, the interested public and indeed some physicists had to seek support for the relativity theory, particularly in the face of its challenging paradoxes and iconoclastic demands, chiefly in the apparent ease with which it explained Michelson's results—an experiment that had counted little if anything for Einstein himself, but which to this day profits from being pedagogically the easiest tool of persuasion (at least in the oversimplified versions found in textbooks). Thus it came about that Einstein's scientific *Weltbild* has been absorbed into the culture of science and beyond with the aid of a rhetoric that had little to do with its genesis.[58]

A decade after his first edition, von Laue published the fourth edition of his successful text, renamed straightforwardly *Die Relativitätstheorie* (1921). By that time, two years after the famous general relativity test of November 1919, most physicists had come to accept Einstein's special relativity over Lorentz's relativistic electrodynamics. Yet, there still was a tendency to confuse them in certain profound respects; and for this reason, von Laue felt compelled to end volume 1 of his 1921 book with a special section, largely taken from his first edition, in which he patiently tried to set the matter straight once more.

The historical sequence of developments, he wrote, had produced the misunderstanding that relativity theory is more closely related to electrodynamics than to mechanics. The source of this misperception is that the transformation equations were indeed first deduced from electrodynamic considerations (by the majestic Lorentz in various, successively better forms, over several years ending by 1905—and most physicists will have absorbed the equations and their original electrodynamic context early in their training). But in modern relativity theory, they applied equally to the phenomena in *all* fields of physics, including mechanics, even though there one usually did not need them to make predictions that are correct to within the error limits of most measurements in mechanics.

A related misunderstanding was, he said, that since all forces of physics are subject to Lorentz transformations, they all may have a common origin, namely, the electrodynamic forces to which Lorentz had first applied them. But that thought, too, was entirely unwarranted. On the contrary, the fact that the relativity principle can be applied equally to all forces hints not at a subordination of mechanics to electrodynamics, but at the "equal subordination of both under higher laws."

While von Laue did not speculate further on the reasons behind the long-lingering confusions, we may point to two that have roots in rhetoric: the "momentum" of the old term "Lorentz transformations," and the long-term persistence of the implication of operational "equivalence" between Lorentz's and Einstein's theories. We are dealing here with what one might call the inertia of rhetoric.

The "equivalence" of two (or more) differently based theories occurs again and again and is one of the surprising facts of science. Famous cases include the equally powerful (for making useful predictions) schemata of Copernicus and his Ptolemaic opponents and the consequences derivable from either Heisenberg's matrix mechanics or Schrödinger's wave mechanics. Richard Feynman has put the puzzle in perspective in a memorable passage on yet another such case, that of the law of gravitation:

> Mathematically each of the three different formulations, Newton's law [of gravitation], the local field method, and the minimum principle, gives exactly the same consequences. What do we do then? You will read in all the books that we cannot decide scientifically on one or the other. That is true. They are equivalent scientifically. It is impossible to make a decision between them if all the consequences are the same. But psychologically they are very different, in two ways. First, philosophically you like them or you do not like them; and training is the only way to beat that disease. Second, psychologically they are very different because they are completely unequivalent when you are trying to guess new laws.[59]

"Guessing new laws" is here shorthand for getting at new science, advancing beyond the stage reached by the different formulations that yielded "equivalent" results on previous puzzles. But—more than in most other endeavors—getting at new science tomorrow is the main purpose of doing science today. So it is a matter of crucial significance that when the two theories are extrapolated beyond the intersection point where, for the needs of the moment, the predictions are (more or less) the same, the next steps on the diverging trajectories are going to be quite different. As we saw in the confrontation between Bohr and his critics, and in the comparison of the Lorentzian and Einsteinian theories, every major theory in science is shaped and propelled by its own list of themata and its own world view. Thereby each sets the stage for a future form of science quite different from its rival—a future stage on which a new cast of characters can present its own acts in the unending play.

I am happy to acknowledge support by a grant from the Andrew W. Mellon Foundation.

APPENDIX

Brief Sketch of Main Oppositions between Some Components of Lorentz' and Einstein's World Pictures*

Some Components of the World Picture	Lorentz Electromagnetic W.P	Einstein Relativistic W.P.
A. goal-purpose	e.m. reduction	logical unification
B. degree of symmetry	ultimate dualism	ultimate monism
C. coherence of concepts and phenomena	logically separate but reducible domains	logically equivalent and synthetic domain
D. consistency of view	determined *a posteriori*	logically entailed
E. focus of interest	basic substances	totality of relations
F. nature of fundamental order	ultimately contingent	transcendental rational necessity
G. theoretical methodology	constructive/inductive	postulational/deductive
H. practical methodology	experimental dialectic	logical critique of concepts
I. attitude towards mathematics	useful tool only	ideal language of reality
J. attitude towards experiments	privileged access to nature	skeptical, epistemological concerns
K. nature of confirmation	mainly empirical evidence	mainly logical perfection
L. nature/value of invariance	appearance, since physical result	axiom, since regulative principle
M. ontology	matter/force/ether	energy-field

*See William Berkson, *Fields of Force: The Development of a World View from Faraday to Einstein* (New York, 1974), p. 254 for a similar graphical approach to the comparison of various scientists' conceptions of nature. I wish to acknowledge the contribution of Mr. Keith Anderton in the preparation of this display.

NOTES

1. Aristotle, *Rhetorica*, in *The Works of Aristotle*, v. XI, W. D. Ross, ed. (Oxford: Clarendon Press, 1924), p. 1356a of Book I.2.

2. Isaac Newton, *Opticks* [1730 edition] (New York: Dover Publications, Inc.), p. 1.

3. *Ibid.*, p. 376. The classic position is summarized in A. Einstein's sentence, "The belief in an external world independent of the perceiving subject is the basis of all natural science." (A. Einstein, *Ideas and Opinions* (New York: Dell Publishing Co., 1954), p. 260).

4. Aristotle, *Rhetorica*, pp. 1357b, 1359a.

5. E.g.,Chapter 2 of G. Holton, *The Scientific Imagination: Case Studies* (Cambridge: Cambridge University Press, 1978), and Chapter 8 of G. Holton, *The Advancement of Science, and Its Burdens* (Cambridge: Cambridge University Press, 1986).

6. Even this brief announcement of my main theme cannot be allowed to pass without a bow to a justly famous work based on analogous (but only analogous) suppositions. I refer, of course, to Alexandre Koyré's *Galilean Studies*, in which he demonstrated that the rhetorical repertoire of Galileo's *Two New Sciences* was divided among five agents/Actors: Salviati, Simplicio, and Sagredo, with their different voices and goals, on center stage throughout; but also, in apparently only brief appearances, "the "Author" (Galileo himself), and most important, the reader/auditor beyond the stage, over whose soul the four others tangle. And then we discover a sixth agent involved in the proceedings—the translators (for example, in the Crew-de Salvio version) who took the liberty of effectively falsifying the text to fit their particular pre-Koyréan epistemology.

   The types of Actors involved and the division of labor between them will be rather different in the cases to be considered here. Still, Koyré's type of analysis has been shown to have relevance even in understanding discussions in contemporary science, e.g., in the study of the unintentionally tape-recorded interactions among three astronomers as they discovered the first optical pulsar, at Steward Observatory on January 16, 1969.

7. Peter Medawar, *The Art of the Soluble* (London: Methuen and Co., Ltd., 1967), p. 7.

8. I shall assume here, rather than repeat or summarize, the elements of the thematic analysis of scientific thought.

9. Michel Foucault, *The Order of Things* (New York: Random House, 1973), p. 159. See also p. xi.

10. E. Segrè, *From X-Rays to Quarks* (New York: W. H. Freeman and Co., 1980), p. 127.

11. L. Rosenfeld, ed., *Niels Bohr, On the Constitution of Atoms and Molecules* (New York: W. A. Benjamin Inc.), p. xli.

12. A. Pais, *Inward Bound* (Oxford: Clarendon Press, 1986), pp. 208–211.

13. A significant but neglected study site is what scientists at the frontier regard as absurd, ugly, unbearable.

14. *Nature* (November 6, 1913), 92:2297, p. 306.

15. N. Bohr, interview of 7 November 1962, p. 1, in American Institute of Physics transcript of *Sources for the History of Quantum Physics* (SHQP).

16. A. Pais, p. 196 (emphasis in original).

17. Rutherford's letter is reprinted in N. Bohr, *Proc. Phys. Soc. 78* (1961), p. 1083, and Bohr's letter in Rosenfeld, *Nils Bohr*, p. xxxviii.

18. J. H. Jeans, "Report on Radiation and the Quantum Theory" (London, *The Electrician*, 1914), p. 89. And not only in the "new mechanics" of the atom—during the first dozen years of the new century, the thema (and hence the rhetoric) of discontinuity had sprung up also in fields as separate as genetics and radioactivity (including "mutation," "transmutation").

19. *Ibid.*, translated from H. Poincaré, "L'Hypotheses des Quanta," in *Dernières Pensées* (Paris: Flammarion, 1913), p. 90.

20. Wassily Kandinsky's autobiographical sketch about the years 1901–1913, in his book *Rückblick* (Baden-Baden: Woldemar Klein Verlag, 1955), p. 16, indicates how he overcame a block in his artistic work at that time: "A scientific event removed the most important obstacle: the further division of the atom. The collapse of the atom model was equivalent, in my soul, to the collapse of the whole world. Suddenly the thickest walls fell. I would not have been amazed if a stone appeared before my eye in the air, melted, and became invisible. Science seemed to me destroyed. . . . "

21. Bohr interview, SHQP, October 31, 1962.

22. *Ibid.*

23. *Ibid.*

24. Quoted in Rosenfeld, *Nils Bohr*, p. xxii.

25. Holton, *Thematic Origins of Scientific Thought* (Cambridge, Mass.: Harvard University Press, revised edition, 1988), p. 156.

26. Quoted in Rosenfeld, *Neils Bohr*, p. xxxiv. Compare Alexander von Humboldt, *Cosmos: A Sketch of a Physical Description of the Universe*, trans. E. C. Otte (London, 1848), vol. I, p. 68: "The charm that exercises the most powerful influence on the mind is derived less from a knowledge of that which is, than from a perception of that which *will be*."

27. " . . . this constant [h] is of such dimension and magnitude that it, together with the mass and charge of the particles [electrons], can determine a length of the order of magnitude required [i.e., for the atom]." Bohr does not even stop to write it down: Radius of the atom $\approx h^2/me^2$.

28. Cf., "The Resistance to 'Reckless' Hypotheses," in *Thematic Origins*, pp. 164–169.

29. *Ibid.*, p. 246.

30. For details, see G. Holton, "More on Mach and Einstein," *Methodology and Science*, 22:2 (1989), pp. 67–81.

31. A significant infelicity in the definition of force was pointed out shortly thereafter by Max Planck. And in the collection of Einstein's own reprints from his desk, given to me by the estate after helping it to organize the archives, there appeared in Einstein's handwriting on the reprint of the 1905 paper a number of corrections to the printed version.

    For documentation of the steps and timetable of Einstein's composition, see John Stachel, ed., *The Collected Papers of Albert Einstein*, vol. 2 (Princeton: Princeton University Press, 1989), pp. 253–274, especially pp. 261–266.

32. A priceless opportunity was thus missed. One can speculate, for example, that in a "revision" Einstein would have shown more clearly the close subterranean connections that have been shown to exist between the relativity paper and the earlier, so-different-appearing papers of 1905, on light emission and Brownian motion. One can also see that the internal structure of the three papers is parallel.

33. As has been done for other purposes in the monograph by Arthur I. Miller, *Albert Einstein's Special Theory of Relativity* (Reading, Mass.: Addison-Wesley Publishing Co., Inc., 1981). Among Miller's other contributions that are useful for the topic of this paper, see especially, "The Physics of Einstein's Relativity Paper. . . . ," *American Journal of Physics 45* (1977), pp. 1040–1048, and "On Einstein's Invention of Special Relativity," in A. I. Miller, *Frontiers of Physics 1900–1911* (Boston: Birkhäuser, 1986), pp. 191–216.

34. Quoted in Holton, *Thematic Origins*, 1988, p. 381. Where not otherwise identified in what follows, brief passages or phrases quoted from Einstein will be found also in that source, Chapters 6–9.

35. Holton, *Thematic Origins*, 1988, p. 382. This early date contradicts a recent speculation that Einstein ascribed great importance to these experiments only when he reflected on the genesis of relativity in his old age.

36. A. Einstein, "Autobiographical Notes," in *Albert Einstein, Philosopher-Scientist*, P. A. Schilpp, ed. (Evanston, Ill.: The Library of Living Philosophers, 1949), p. 27. Miller, in *Frontiers*, p. 200, documents that by 1905 Einstein's "investigations of the structure of light revealed that classical electromagnetism failed in volumes of the order of the electron's. Thus, the electromagnetic worldpicture could not succeed. His Brownian motion investigations had yielded a similar result for mechanics; hence, exit any possibility for a mechanical worldpicture." Additional support comes from Einstein's letters, e.g., to M. von Laue and C. Seelig. In one of the early, neglected statements, a letter to E. Bovet of 7 June 1922 (*Wissen und Leben*, vol. 24 (1922), p. 902), Einstein refers to his theory as "an improvement and modification of the basis of the physical-causal worldpicture."

37. H. A. Lorentz, "Electrodynamic Phenomena. . . . ," *Proceedings of the Royal Academy of Amsterdam, 6* (1904), p. 809.

38. Miller, *Special Theory*, Section 1.15.1, has a useful discussion of the sources on which Einstein drew ("definitely," "very probably," "maybe").

39. Quoted in Holton, *Thematic Origins*, p. 253.

40. M. Planck, "Das Prinzip der Relativität. . . . ," *Verh. D. Phys. Ges. 4* (1906), pp. 136–141.

41. M. Planck, *"Die Kaufmannsche Messung. . . . , Phys. Z., 7* (1906), pp. 753–761.

42. H. A. Lorentz, *Versuch einer Theorie* (Leiden: Brill, 1985).

43. Quoted in Holton, *Thematic Origins*, p. 310.

44. Lorentz, "Electromagnetic Phenomena." See also *Versuch einer Theorie*, in which Lorentz calls one of his own hypotheses initially "estranging" (p. 123), and comments on another one that "to be sure, there is no basis for it" (p. 124).

45. Quoted in Holton, *Thematic Origins*, p. 321.

46. In the *Appendix* we sketch briefly some of the main oppositions between the two world pictures to show at a glance how fundamentally they differ.

47. These papers have been analyzed well by Miller, *Special Theory*, pp. 225–235, 333–352, and in his *Frontiers*, Essay 1.

48. W. Kaufmann, "Über die Konstitution. . . . ," *Annalen der Physik 19* (1906), pp. 487–553.

49. W. Kaufmann, "Nachtrag. . . . ," *Annalen der Physik 20* (1906), pp. 639–640.

50. For the Lorentz-Poincaré responses, see Miller, *Special Theory*, pp. 334–337.

51. A. Einstein, *"Relativitätsprinzip. . . . ," Jahrbuch Radioact. 4* (1907), pp. 411–462.
    In a brief article dated August 1906 (A. Einstein, "Über eine Methode. . . . ," *Annalen der Physik 21* (1906), pp. 583–586), Einstein had mentioned Kaufmann's experiment with β-rays in passing, but studiously avoided discussing it, commenting on its results or worth, or even giving the bibliographic reference to it. Instead, Einstein proposed a different and better experiment to test the rival theories, one using slow cathode rays (apparently unaware that Planck had recently made a similar suggestion). Einstein predicted significant differences to result that would distinguish among three theories, namely, "Theorie von Bucherer," "Theorie von Abraham," and "Theorie von Lorentz und Einstein." For the last, we note the use of the singular and of his own name in his text—all quite uncharacteristic for him. For this brief piece only, Einstein adopted or aped Kaufmann's terminology, and probably did so with tongue in cheek.

52. Planck, *"Das Prinzip der Relativität. . . . "*

53. Planck, *"Die Kaufmannsche Messung. . . . "*

54. *Phys Zs. 7* (1906), pp. 759–761.

55. Quoted in Holton, *The Scientific Imagination*, p. 10.

56. W. Wien, *Über Elektronen*, Second Edition (Leipzig: Teubner, 1909), p. 32.

57. (Braunschweig: F. Vieweg & Sohn, 1911). See especially pp. 18–21.

58. A whole study on the Rhetoric of Appropriation/Rejection could be devoted to the events subsequent to the publication in 1969 of the results of an historical study of the documents that showed we must take Einstein at his word that the genetic influence of the Michelson experiments was at best indirect and small. This finding contradicted the almost universal agreement at the time among historians and philosophers of science, as well as textbook writers, who held that the experiments had been a crucial guide for Einstein. That view was part of, and reinforced, a branch of the philosophy of science that incorporated experimenticism.

   Since that publication in 1969 (see Holton, *Thematic Origins*, Chapter 8 and pp. 477–480) all additional, first-hand documents have supported the conclusion reached in 1969. Moreover, by and large it has been fully incorporated into the current view. But even though pedagogic texts now have to deprive themselves of the simple way to make plausible the origins and necessity of relativity, many of them still feel required to go through the motions, hoping thereby to persuade their students more easily about this demanding and counterintuitive theory. Thus the authors of *The Story of Physics*, Lloyd Motz and Jefferson Hane Weaver (New York: Plenum Press, 1989), pp. 252–253: "Physics [about 1900] was growing rapidly . . . but one experiment brought consternation and confusion—the famous Michelson-Morley experiment. Since this experiment had a direct bearing on the acceptance of the theory of relativity, *although historical evidence indicates that Einstein did not know about it when he wrote his relativity paper*, it is useful to examine this experiment" (italics supplied). And they proceed to do so at length.

59. Richard Feynman, *The Character of Physical Law* (Cambridge, Mass./London, U.K.: MIT Press, 1967), p. 53. I thank S. Sigurdsson for having drawn my attention to this passage.

Notes on Contributors

GERALD HOLTON is Mallinckrodt Professor of Physics and Professor of History of Science at Harvard University, and Visiting Professor at Massachusetts Institute of Technology. His researches in the history of science have focused on modern physics, particularly on the rise of relativity. His books include: *Thematic Origins of Scientific Thought: Kepler to Einstein* (2nd edition, 1988), *The Scientific Imagination: Case Studies* (Cambridge University Press, 1978), and *The Advancement of Science, and its Burdens: The Jefferson Lecture and Other Essays* (Cambridge University Press, 1986).

PHILIP KITCHER received his B.A. from Cambridge University and his Ph.D. from Princeton University. His is Professor of Philosophy and Faculty Coordinator for Science Studies at the University of California at San Diego. His is the author of *Abusing Science: The Case Against Creationism* (MIT Press, 1982), *The Nature of Mathematical Knowledge* (Oxford University Press, 1983), *Vaulting Ambition: Sociobiology and the Quest for Human Nature* (MIT Press, 1985.) With William Aspray he is co-editor of *History and Philosophy of Modern Mathematics* (Minnesota Studies in the Philosophy of Science, Volume XI, 1987), and with Wesley Salmon he is co-editor of *Scientific Explanation* (Minnesota Studies in Philosophy of Science, Volume XIII, 1989). He is currently at work on a book on progress and rationality in science.

PETER MACHAMER received his B.A. from Columbia University, his M.A. from Cambridge University, and his Ph.D. from the University of Chicago. He is Professor and Chairman, Department of History and Philosophy of Science at the University of Pittsburgh. He works on many topics in 17th Century science and philosophy. Most recently, he contributed a chapter on the philosophy of psychology for the department's jointly written new text, *Introduction to Philosophy of Science* (Prentice-Hall).

MAURIZIO MAMIANI is Professor of History of Science and Technology at the University of Udine (Italy). He is the author of several books and papers about Isaac Newton's natural philosophy including *Il prisma di Newton* (Bari: Laterza, 1986) and Introduzione a Newton (Bari: Laterza, 1990). His research interests include Cartesian philosophy and its influence on the modern concept of space, Teorie dello spazio da Descartes a Newton (Milano: Franco Angeli, 1981) and the origin of the modern encyclopedias in La Mappa del sapere :*La classificazione delle scienze nella Cyclopaedia di E. Chambors* (Milano: Franco Angeli, 1983). Current projects include an edition of some of Newton's theological manuscripts. His most recent paper concerning Newton's intended appendix to the second edition of Principia (1713) will shortly appear in *Nuncius*.

ERNAN MCMULLIN is Director of the Program in History and Philosophy of Science and O'Hara Professor of Philosophy at the University of Notre Dame. He is past President of the Philosophy of Science Association, the American Philosophical Association (Central Division), the American Catholic Philosophical Association, and the Metaphysical Society of America. He is a Fellow of the American Academy of Arts and Sciences, of the American Association for the Advancement of Science, and of the International Academy for the History of Science. His books include: *Newton on Matter and Activity* (1978), *The Concept of Matter*, ed.(1963), *Galileo, Man of Science*, ed. (1967), *Evolution and Creation,* ed. (1985), *Construction and Constraint,* ed. (1988), *The Philosophical Consequences of Quantum Theory*, ed. with J. Cushing (1989), and *The Social Dimensions of Science,* ed. (1991). He is currently working on a book tentatively titled: *Rationality, Realism, and the Growth of Knowledge.*

MARCELLO PERA taught philosophy of science at the University of Pisa before being appointed to a Chair of Philosophy at the University of Catania in Sicily. He is a Fellow of the Center for the Philosophy of Science at the University of Pittsburgh and he has lectured extensively in American and Canadian universities. He is author of numerous publications, including *Apologia del metodo; Induzione e Metodo Scientifico; Popper e la scienza su palafitte; Hume, Kant e l'induzione;* and *The Ambiguous Frog: The Galvani-Volta Controversy on Animal Electricity.* His latest book, *Science and Rhetoric*, which appeared in Italian in 1991, will shortly be published in English.

PAOLO ROSSI was born in Urbino (Italy) in 1923. He is Professor of the History of Philosophy at the University of Florence. Many of his books have been translated into Spanish, Japanese, and Polish. Among those translated into English are: *Francis Bacon from Magic to Science* (University of Chicago Press, 1968), *Philosophy, Technology and the Arts in the Early Modern Era* (Harper, 1970), *The Dark Abyss of Time: The History of the Earth and the History of Nations from Hooke to Vico (University of Chicago Press,* 1984 and 1987). *Clavis universalis : arti della memoria e logica combinatoria da Lullo a Leibniz,* first published in 1960, has been reissued by Il Mulino (Bologna) in 1984. In 1985, Professor Rossi was awarded the Sarton Medal of the History of Science Society. He is President of the Florence Center for the History and Philosophy of Science and a Fellow of the Accademia Nazionale dei Lincei.

DUDLEY SHAPERE received his B.A., M.A., and Ph.D. degrees from Harvard University. Prior to going to Wake Forest University where he is Z.Smith Reynolds Professor of the Philosophy and History of Science, he taught at Ohio State University, The University of Illinois, The University of Chicago, and the University of Maryland.

He received the Quantrell Award for Excellence in Undergraduate Teaching at the University of Chicago, and the Distinguished Scholar-Teacher Award at the University of Maryland. From 1966 to 1975 he served as Special Consultant (Program Director) for the History and Philosophy of Science Program at the National Science Foundation. Among his forthcoming publications are "The Universe of Modern Science and Its

Philosophical Exploration" (Philosophy and the *Origin and Evolution of the Universe,* E. Agazzi and A. Cordero, eds.), and "The Origin and Nature of Metaphysics" *(Philosophical Topics),* "On the Introduction of New Ideas in Science"*(The Creation of Ideas in Science,* J. Leplin, ed.).

WILLIAM R. SHEA is Professor of History and Philosophy of Science, and a member of the Centre for Medicine, Ethics and Law at McGill University, Montreal. A graduate of the University of Cambridge, and a former Fellow of Harvard University and the Institute for Advanced Study in Berlin, Professor Shea is President of the International Union of History and Philosophy of Science, and a Fellow of the Royal Society of Canada. He is the author of G*alileo's Intellectual Revolution* and *Copernico, Galileo, Cartesio: Aspetti della Rivoluzione Scientifica* and the editor of several books including *Reason, Experiment and Mysticism in the Scientific Revolution,* and *Scientists and their Responsibility.* His latest book, *The Magic of Numbers and Motion: the Scientific Career of René Descartes,* was published by Science History Publications/USA in 1991.

RICHARD S. WESTFALL, Professor Emeritus of the History of Science at Indiana University, has devoted his professional life to the period of the scientific revolution, which includes Galileo and Newton. He is the author of a number of articles on Newton together with two books: *Force in Newton's Physics: The Science of Dynamics in the Seventeenth Century* (1971) and *Never at Rest: A Biography of Isaac Newton* (1980). Recently he published a volume on Galileo: *Essays on the Trial of Galileo* (1989). He is currently at work on a social history of the scientific community during that era.

Index of Names

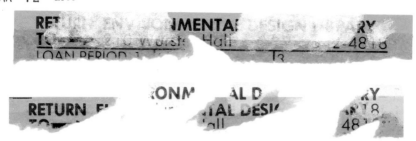